Beekeeping Blueprint

Beekeeping Blueprint

Start Your First Hive,
Raise Healthy Bees And
Sustainably Harvest Honey
Year After Year

A Complete Guide To Backyard
Beekeeping For Beginners

Jason Chrisman

For my brother Jarrod

Table Of Contents

III. Becoming a Seasoned Beekeeper

My Beekeeping Story

Bees literally flew into my life on a warm spring day in 2009, with temperatures ranging in the mid-80s. I'd been working on a roofing job, and a colleague had just dropped me home. As I reached for my lunchbox at the back of his truck, he yelled, "Jason! You've got a problem, bud! Look up over there, under that oak tree."

I went around to see what he was talking about. Pointing to a huge swarm of bees, he said, "You're going to have to call a beekeeper." Sure enough, we could see a steady traffic of bees flying to and from the swarm. We then moved under the tree's canopy, which revealed something slightly larger than the size of a basketball. I couldn't help but notice how organized and calm the bees were as they moved about.

After taking it all in, I finally said, "I think it's time for *me* to become a beekeeper." And that's what I did.

My grandfather kept bees, but I mostly observed from a distance and had no real experience with them at that point in my life. Luckily, I knew a guy who knew a guy—my neighbor Tim had friends who kept bees. I gave him a call that evening.

A couple of winters before that fateful day, Mark and I were out riding our ATVs. We stopped by a creek, and as we watched the water flow by, I shared that a part of me had always wanted to be a beekeeper. When that swarm arrived in a tree not thirty yards from my house, I knew that time had come.

Tim (who, by the way, knew nothing about beekeeping himself but was willing to join in on the fun) was able to source a box into which we could transfer the swarm. The next day, he came over to help move the bees. A true beekeeper's approach is to cut the limb off first, take it down to the ground, and then gently shake all the bees into the prepared box. Knowing what I know now, it's important to hold the tree limb with one hand so it doesn't go crashing to the ground. The other hand is working the saw while also helping to keep balance on the ladder. That's how it's supposed to be done.

In our case, we were about ten feet in the air performing an acrobatic feat with hundreds of bees flying around us. I was sure that I'd be stung at any moment. Tim and I cut off the tree limb, dropped it into the box, and then ran away as fast as we could, screaming.

The arrival of those bees changed my life. At the time, my daughter was around six, so neither my wife nor I wanted a bee operation set up too close to home. Although we did live on five acres of land and had the space to set up hives far enough, I came across some uninhabited land down the road where there were more options for nectar. (The closer bees are to the source, the faster their hives get packed with honey.) I contacted the land-owner and asked if I could place some hives there.

During this early period, I set up my JC's Bees YouTube channel as a personal diary of sorts. I shared everything—my successes and many failures—and I was shocked at the number of people who embraced my story. The online beekeeping community cheered me on and offered advice when I ran into issues. I never expected tens of thousands of people to be interested in following me on my journey, but I must say that I am humbled and honored.

If I could just pause here for a moment and rewind to six years before that swarm serendipitously showed up in my yard, I'd like to share that I had been going through one of the most difficult challenges in my life. In 2003, my brother Jarrod was tragically shot and killed. He and I were very close and did everything together. Jarrod was and still is one of the dearest people in the world to me. Dealing with his death weighed heavily.

Adding to the weight, my family and I watched and listened as lawyers described photos from the crime scene in detail after the murderer was caught. Grieving for a loved one is hard enough—throw the harrowing court proceedings into the mix, and it can certainly take an emotional toll that requires months, years even, before the healing process can begin.

With every passing day, Jarrod's loss still felt fresh. I may have appeared just fine on the outside, but I continued to move through life with a heavy weight on my shoulders.

Coming full circle, the arrival of that swarm is what helped me through this dark time. The challenges and joys of beekeeping soothed my grief, and my fascination with bees quieted my mind. As I got lost in my hives, the heaviness slowly began to lift. Of course, I still miss my brother today, but beekeeping has granted me a measure of peace.

Fast forward to the summer of 2022, I asked JC's Bees subscribers to share their beekeeping "Why"s. The quotes you'll find throughout the book mirror some of these personal responses. While each of our stories is unique, beekeepers from across the country and the world confirm the incredible benefits bees have brought them.

When I unexpectedly found that swarm on my property, I couldn't have imagined becoming such an invested beekeeper now taking care of over 30 colonies. Over a decade into the practice, I'm privileged to take the knowledge I've gained over the years and package it into this book so even more people can discover the joys of working with bees. I equally hope it will help readers avoid some of the mistakes I made in my early days. *Are you ready to embark on an adventure of a lifetime?*

Introduction

Being with my bees out in the open air, an indescribable feeling of peace and calm takes over like nothing I've ever felt. I consider myself lucky to be able to tend to my hives and make a positive return at the same time. There are ways (including a few you might not expect) to make beekeeping pay for itself—it's just a matter of sticking around long enough to find out how.

In *Beekeeping Blueprint*, you'll learn everything you need to set up and manage your first hives. The book is split into three parts:

1. Learning the Essentials
2. Building Momentum
3. Becoming a Seasoned Beekeeper

Each section is organized to accompany you as you slowly increase your knowledge and develop your beekeeping skills.

Before Diving In

One of the most important points I want to drive home before going any further is that beekeeping is not just a hobby to be picked up in your free time and shoved to the side when things get busy. In the 1980s, that might have been possible, but bees are threatened by far too many pests these days for them not to be managed regularly. If you plan on buying a hive, leaving it in the corner of your yard and staring at it from the kitchen window, you're better off saving your money and putting it into something else!

With that settled, here are a few practical considerations I wish people had told me before diving into the world of beekeeping:

1. Working with nature can be unpredictable. Expect to suffer some setbacks along the way. The ups and downs are all part of the game, so don't give up.
2. As you'll be reminded throughout the book, bees need consistent care and management if they're going to thrive in a place of your choosing (as opposed to their natural habitat). Generally, you'll need to spend at least 30–45 minutes every 10 to 12 days per hive, with extra time in the spring. If you don't take an active role, your bees won't make it.
3. Work out your budget. While picking up secondhand items can save you some cash, it's best to budget around $600–$800 to get set up. As a bare minimum, you'll need to purchase a hive, bees, a hive tool, smoker, bee suit and gloves—more on each of those in Chapter 4.
4. Know and understand your climate. Bees have evolved to work in harmony with the natural availability of pollen and nectar. Knowing when flowers bloom in your area will help support your colony. Higher temperatures resulting from climate change have caused flowers to bloom earlier than usual or for a shorter period. Persistent heat waves can also be harmful to bees.

Since I live in the Midwest, the information in this book should be supplemented with research on your own area. Consider joining a local chapter of beekeepers in person or online.

5. Research the legality of beekeeping in your neighborhood, as it varies. In San Diego, California, for example, the city stipulates that apiaries must be at least 15 feet from your property line. They also require a six-foot tall barrier to surround the beehive unless it's at least eight feet above the ground. More information is available on the *Apiary Inspectors of America* website.[1]

1   https://apiaryinspectors.org/state-laws/

6. Inform neighbours that you plan on keeping bees, as this allows you to address any concerns and provide reassurance. If they're still worried, you might want to place your hives further away from your property line. On the other hand, they might love the idea of having bees next door and enjoying some fresh honey— you won't know until you ask!

7. Commit to learning. Before setting up your hive, do your research. If you're reading this book, you're on the right track! You can also check out the JC's Bees YouTube channel and Beekeeping Blueprint Community to stay in the loop. Local beekeeping chapters can equally provide advice specific to your location.

8. No matter how much you think you know about bees, they always have something new to teach. Heck, after all these years, I'm still learning, and I know it's a journey that'll never end.

As you follow the steps outlined in this book and begin developing your beekeeping skills, there's a high chance you'll better connect with and understand nature. While you're at it, you'll also be doing your own little bit to help preserve the environment for generations to come. I'm so excited to share more about this traditional practice with you. Let's get started!

PART I

Learning The Essentials

"Many years ago, a friend asked if I wanted to see his bees. I said, 'No way!' and wanted nothing to do with those stinging bugs. Years later, my wife and I noticed our garden wasn't doing very well. We thought having bees on our property would not only help the garden but also pollinate other things around our area. I dove into beekeeping and have loved it ever since. I am always learning something new."

—Rob Alfred

CHAPTER 1

The "Why"

Every beekeeper has their own story about how and why they ventured into the world of bees. If you're reading this book, something has most likely brought you here. While I'm always thrilled to learn about people's unique stories, there are underlying factors about bees and beekeeping that bring us all together.

Before rushing out and spending any money, the best advice I can give about starting your beekeeping journey is to know what you're getting yourself into and *why*. If that's not entirely clear to you yet, keep reading.

Dealing With Royalty

After Queen Elizabeth's passing in September 2022, John Chapple, the royal family's official beekeeper, placed a black ribbon on each of Buckingham Palace's five hives and gently shared the solemn news. With every colony holding around 20,000 bees, he knocked on each one and told the bees not to go. King Charles III, he assured, would be a good master to them.[2]

In the world of apiarists, "telling the bees" is an age-old tradition. Possibly dating back to the ancient Greeks, who believed that bees bridged our world with the afterlife, not informing bees about a new owner could lead to them stopping honey production, abandoning the hive or dying. While I can't say the notion of telling the bees all your family news is backed by any hard science, I

2 https://www.independent.co.uk/life-style/royal-family/royal-beekeeper-bees-queen-death-b2164345.html

Telling the bees: an age-old tradition
informing bees of their new owner

fruit from a poorly
pollinated flower

fruit from a
well-pollinated flower

wholeheartedly agree that the status of bees on our planet deserves royal consideration.

Bees matter. They help produce a third of the world's food or one in every three mouthfuls we eat. Losing honeybees can have far-reaching consequences. For example, almond trees need honeybees to pollinate their flowers. No bees, no almonds. As a result, there's been an increased demand for bees in states such as California, the largest almond producer in the U.S.[3]

While different insects can pollinate plants, bees are the most talented since they always pollinate the whole flower, giving perfect fruit. When you get a strange-looking fruit from the store, it could be that another insect pollinated it and didn't do as good a job as a bee would have, or that the fruit tree's flowers were visited less often.

Perfect fruit or not, understanding the impact bees have across the ecosystem of our everyday lives, we can hopefully all agree they deserve our utmost respect.

3 https://www.agmrc.org/commodities-products/nuts/almonds

Joining the Cause

Mainly due to human activity, bees are at risk of extinction with modern agriculture wiping out acres of plants that provide the nectar and pollen that honeybees need. Millions of hectares of precious woodland have also been cut down, where honeybees would otherwise make homes in the hollows of trees. When we consider environmental factors, such as invasive species, the destruction of habitats, climate change and pesticides, it's no wonder bees are struggling.

In China, since honeybees are in sharp decline due to pollution and pesticides, farms pay workers to hand pollinate their crops—a far more expensive and less effective method than what honeybees offer!

In 2008, Colony Collapse Disorder (CCD) was directly responsible for a 60% reduction in beehive populations since the late 1940s. Back then, there were roughly 6 million active beehives in the United States. By 2008, this number had dropped to just 2.4 million.[4]

4   https://www.beepods.com/why-beekeeping-has-become-so-popular-and-why-its-time-you-jumped-on-the-bandwagon/

The 4 Ps

parasites

pathogens

pesticides

poor nutrition

There are various theories behind the cause of CCD. One of these points to the 4Ps: parasites, poor nutrition, pesticides and pathogens. Modern farming methods are destroying natural pollinator habitats and removing areas of diverse flora and fauna, replacing them with monocultures. Bees are then left with little choice and resort to setting up shop in people's houses, barns or chimneys. This biodiversity loss also negatively impacts the variety in their diet, leading to a lack of nutrients. As a result, bees are prone to pathogens that cause disease, opening the way for parasites (including my personal nemesis, the Varroa mite).

Moreover, when bees feed on pesticides, they take them back to their hive, contaminating their food (and our honey). This contamination can lead to birth defects in new bees. Before long, such hives are depleted of healthy bees, and the entire colony eventually dies out.

Upon discovering that bees could be kept in small hives in urban areas, city dwellers have developed an interest in beekeeping. Urban bees actually tend to be healthier than rural bees. Not only are urban bees able to find plenty of food from weeds, flower boxes and herbs, but they're also kept safe from the pesticides used on crops and monocultures.

If you're in the starting phases of beekeeping, you're right on time. Whether you're based in a rural or urban setting, creating a safe, flourishing environment for bees has a ripple effect that positively impacts the natural ecosystem, including our own health and well-being.

Personal Benefits of Beekeeping

While environmental altruism is more than enough to draw people to beekeeping, another popular motivation is recognizing the

Harvesting honey and beeswax are some of the many benefits of beekeeping

benefits bees bring to your garden. In addition, who wouldn't mind access to free honey or other bee products such as beeswax, pollen, royal jelly and propolis?

Did you know that eating local honey boosts the immune system and can be a natural way to deal with seasonal allergies?

What's more, becoming a beekeeper offers personal advantages that go beyond the obvious: a greater connection with nature, better mental health and access to a supportive community.

Connection With Nature

There's a lot to be learned by simply spending time with bees and observing them as they go about their business. Since the behavior and health of a colony are tied to the seasons and environment, you'll naturally become more in tune with the elements. When your bees are flying off to harvest nectar, you'll know that there's a good crop nearby. You'll also be able to sense your bees' mood and figure out when they're feeling cheerful and productive or if they're feeling under the weather because of drought or disease— all indicators that it's time for you to step in and give them a hand.

If you have children, bees can educate them about the crucial role insects play in the food chain and how we can all enjoy a brighter future by protecting bees.

Mental Health

Terrifying stinging insects to most, beekeepers stand among their bees in serenity and awe, recognizing the role they play in connecting us to Mother Nature. To me, there's nothing more calming than tending to my bees as they fly all around me.

Once an apiarist works through the initial fears of getting stung, it's a common experience to enter a near trance watching bees at work. Call it meditation, mindfulness or whatever you want—beekeeping is healing.

Dealing with the traumatic loss of my brother in 2003, I know I'm not the only one who has experienced these healing effects. Scientific research confirms the benefits of beekeeping for those with PTSD. For any veterans reading, *Hives for Heroes* is one of the many organizations that promote beekeeping to support a healthy transition from service.[5]

Bees remind us to slow down. They will generally coexist peacefully with us when we remain relaxed and calm, keeping

5   https://hivesforheroes.com/

movements slow and gentle. When we show signs of agitation, they too are more likely to get all worked up. Overall, bees teach us to find harmony with ourselves and each other.

A Buzzing Community

When the plight of bees became public knowledge around 2008, people rose to the occasion, and fortunately, beekeeping has since exploded in popularity. Beekeepers provide bees with a vital service and support them in doing what they do best. They give bees somewhere to live, feed bees when needed, treat them for parasites and help the colony grow.

In the U.S. alone, there are over 100,000 beekeepers.

Whatever has inspired your interest in caring for bees, know that you are joining a thriving community of people who are each doing their part to help protect these important little creatures. A great place to start is by reaching out to county or city beekeeping chapters that often host in-person and virtual meetups. These local groups are also a great way to find mentorship opportunities and join the discussion on local conditions relevant to bees and beekeeping.

I'll also add that from what started as a personal diary of my beekeeping journey, the *JC's Bees* YouTube channel has now grown to a community of close to 60K individuals. This number continues to grow steadily. Through my content, I love connecting with other beekeepers, inspiring viewers to perfect their skills and finding inspiration from other community members.

The beekeeping community is a special one. Many have said that they found *their people* when they joined the ranks of beekeepers. As you become initiated into this traditional practice, my wish is that you'll find the same sense of satisfaction and belonging.

"A friend of mine started beekeeping first, and that was all he talked about. I wanted something to do when I retired, so I got two hives—I was hooked. That was 12 years ago. I'm now almost 70 and have 45 hives. I tell folks, 'If you don't believe in God, get some bees.' They teach me something every day. I become the calmest I am when I'm in the middle of 60,000 bees. It's hard to think about anything else."

—Don Stanley

CHAPTER 2
Bee Basics

For the aspiring beekeeper, it's helpful to understand the very basics of how these little critters are built and operate. If you're the type who prefers to just jump right in and figure things out as you go, then by all means, but know that some of what you'll learn in this chapter might save you lots of time, money and head scratching.

The more you know about bees, the better you will be able to connect and work with them. As we go over a bee's anatomy and lifecycle, in addition to different bee breeds, I can guarantee that you'll learn a few things that will take you by surprise, cultivating an even greater appreciation for these marvelous creatures.

Anatomy

Bees may be tiny, but their bodies are complex and rather sophisticated. While bee anatomy is generally the same all around, worker, drone and queen bees have certain variations that reflect the differences in their role.

Whereas humans and other animals have an endoskeleton (or internal skeleton), bees, like many other insects, are equipped with the following:

- an exoskeleton (or external skeleton) made up of three main parts: the head, thorax and abdomen
- two antennae
- three pairs of legs
- two pairs of wings

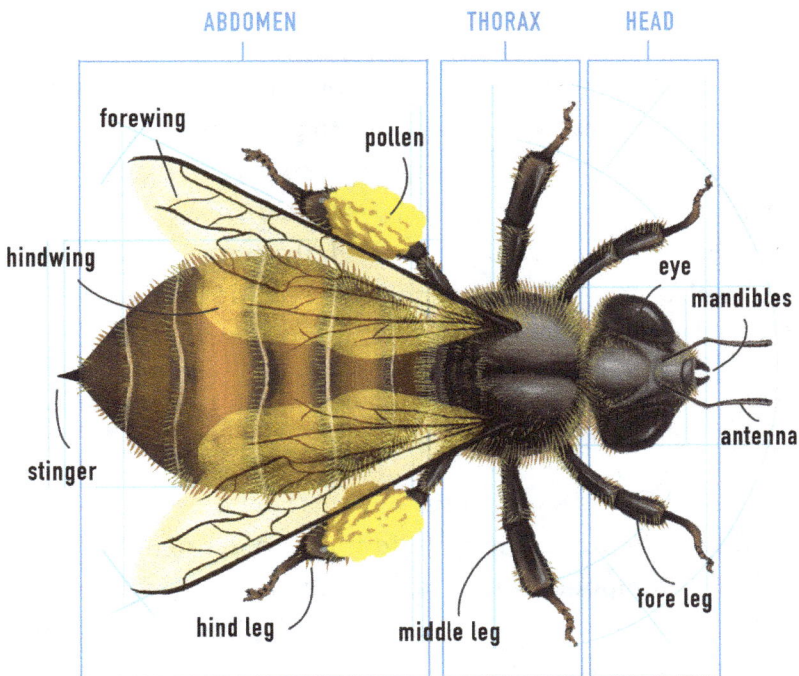

Honeybee anatomy

The Head

ANTENNAE

Bees don't have ears in the same way humans do. Instead, they have two antennae, each formed by an elbow-like joint. These antennae have **mechanoreceptors** that allow bees to hear and feel their surroundings. Through touch, the antennae equally serve to communicate with other bees. The antennae also give bees their sense of smell and taste.

Bees usually rely on the right antenna to get around. Studies have shown that bees don't operate as effectively using their left antenna. It's kind of like our human tendency to be either right- or left-handed.

EYES

A bee's two large eyes are called its compound eyes, made up of many smaller eyes, which each take in a separate image. These images are then sent to the brain and processed into a single image.

Bees also have three simple eyes (ocelli), each containing a single lens to collect UV light. Through the simple eyes, pollen is perceived as dark spots, making it easier to find food. Having a polarized vision allows bees to navigate and process information, as well as protects the eyes from the sun's rays.

MOUTH

A bee's mouth contains salivary glands that do several things, which include producing a liquid that can dissolve sugar and producing compounds used to clean the body. This liquid also creates the hive's **chemical identity**, which gives off a scent that allows bees from the same colony to recognize each other.

Although not found in the mouth, the bee's hypopharyngeal glands produce secretions along with the salivary glands to make royal jelly. The hypopharyngeal glands are also responsible for making wax.

drone

queen

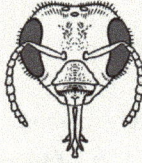

worker

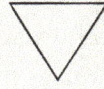
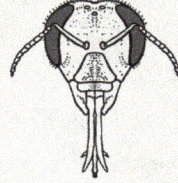

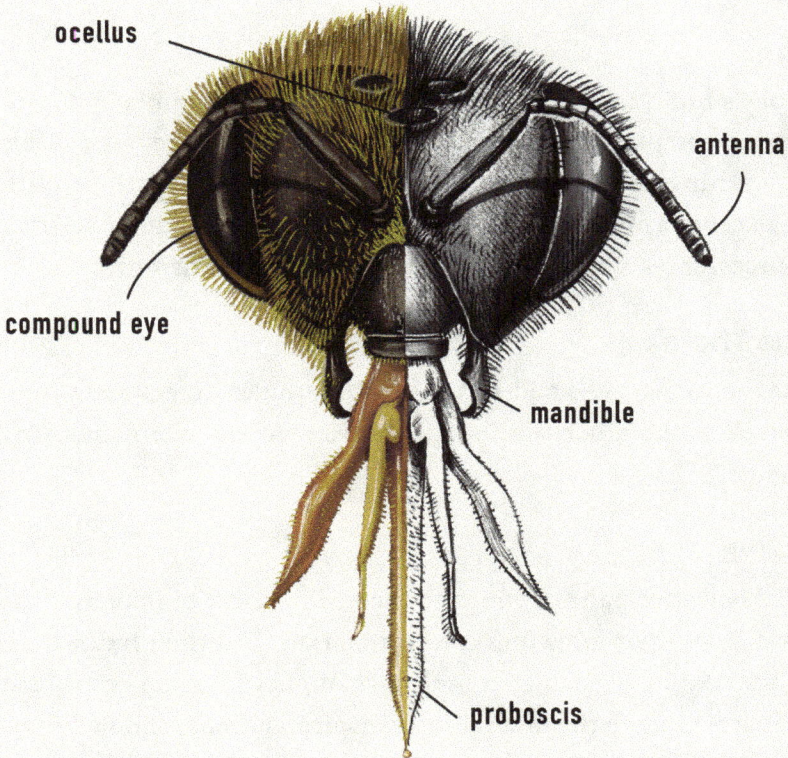

ocellus

antenna

compound eye

mandible

proboscis

A closer look at a honeybee's head

The proboscis is the bee's tongue. It's soft and extendable like a human tongue, although much longer than ours, allowing it to reach that juicy nectar hidden at the center of flowers. The proboscis is also used for grooming.

JAW

If we look at the slight differences that exist in bee anatomy, worker bees have a smooth set of jaws used to make wax. Queen and drone bees have pointed jaws, making it easier to cut and bite.

BRAIN

A bee's brain might be small, but it has an incredible ability to process complex information and make decisions. Bees are capable of learning and retaining a vast amount of information! With an excellent memory, they can remember the location of nectar sources, spreading the word once they return to their hive.

The Thorax

The thorax is a bee's midsection, which includes its legs and wings, as well as the muscles that allow it to control its wings during flight.

WINGS

Bees have two pairs of wings connected by a row of hooks on the hindwing. Their forewings are much larger than their hindwings. In the 1930s, French entomologist August Magnan concluded that according to the mechanics of an airplane, bees should not be able to fly. However, unlike airplanes, a bee's wings rotate to create pockets of low air pressure (almost like mini hurricanes), lifting and propelling them through the air at around 15 miles per hour.

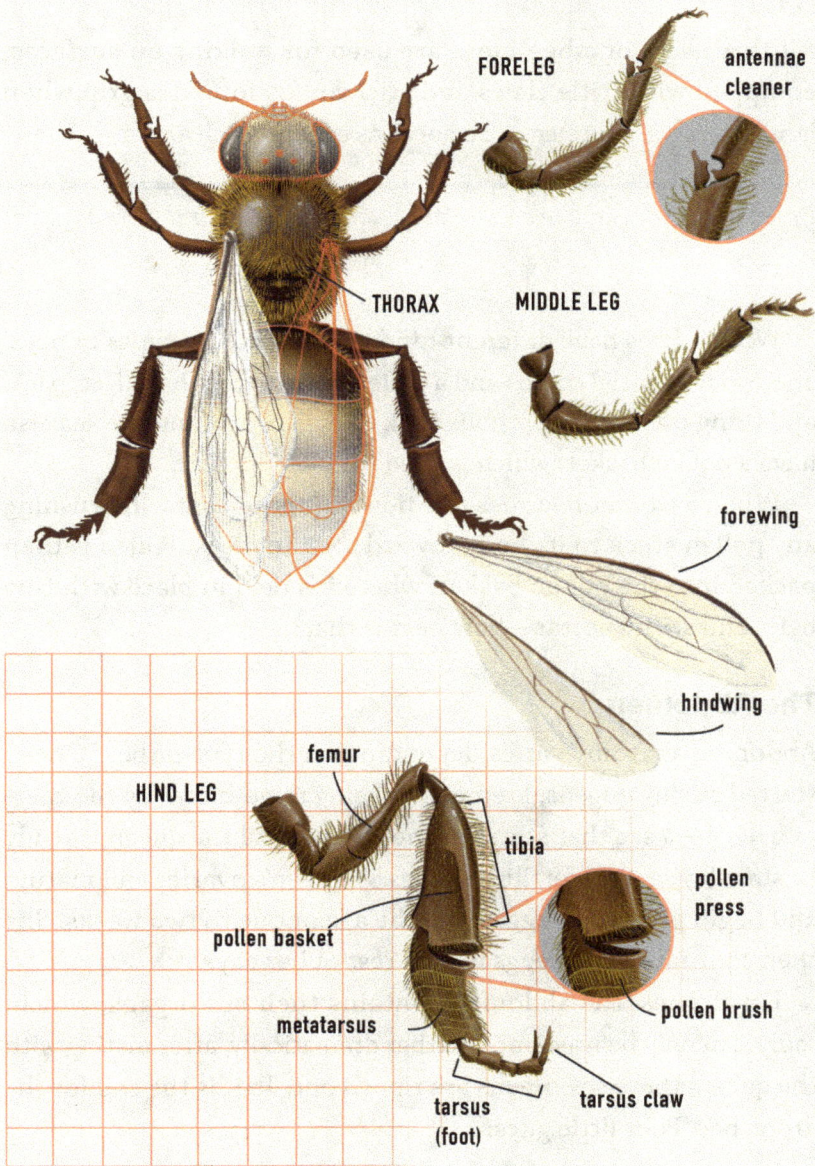

FORELEG

antennae
cleaner

THORAX

MIDDLE LEG

forewing

hindwing

femur

HIND LEG

tibia

pollen
press

pollen basket

pollen brush

metatarsus

tarsus claw

tarsus
(foot)

A bee's thorax includes its legs, wings and the muscles used for flying

LEGS

All three sets of a bee's legs are used for walking on surfaces, equipped with little claws and sticky pads for extra grip when landing. A bee's forelegs also serve as antennae cleaners.

Believe it or not, a bee's taste receptors can be found at the tips of its legs!

Worker bees have different hind legs from other types of bees. Theirs have special combs and a pollen press to brush, collect, pack, and bring pollen and propolis back to the hive. Many species also have a pollen basket, which is used to collect pollen.

When a worker bee goes to a flower, it grooms itself, brushing any pollen stuck to its body toward the hind legs. Pollen is then packed into the pollen basket, where it's held in place with tiny hairs and sticky nectar—how neat is that?

The Abdomen

Abdomen anatomy varies depending on the type of bee. Once I started raising queens, I remember being mind-blown by the **spermatheca**—a sac that stores drone sperm until the queen is ready to start laying and fertilizing eggs. A queen's ovaries will mature and begin producing eggs from the age of one to two weeks. She then continues to lay eggs for the rest of her days.

For drones, the abdomen contains their sex organs, which, sadly, can only be used once. Rather dramatically, after mating with the queen, these sex organs are ripped out. That is the end for the drone bee. Poor little guys.

A drone bee's ejaculation is so loud that humans can hear it. Seriously, it makes a slight popping sound!

As for worker bees, the abdomen's underside contains four pairs of wax glands. The glands secrete liquid wax that hardens when in contact with the air, allowing young worker bees to create wax scales in the hive. About six days later, the wax gland degenerates and these worker bees move on to other roles as they can no longer produce wax.

The degeneration of the wax gland partly explains why the queen is constantly laying eggs. To maintain the size of the colony, enough adolescent workers are needed to keep up with wax production. The wax is used to make comb for raising brood and storing honey and pollen for winter.

STINGER

The stinger is probably the part of the bee people think most about. As you enter the world of beekeeping, take comfort in knowing that bees only use their stinger as a last resort. They don't want to sting you any more than you want to be stung!

worker   **queen**

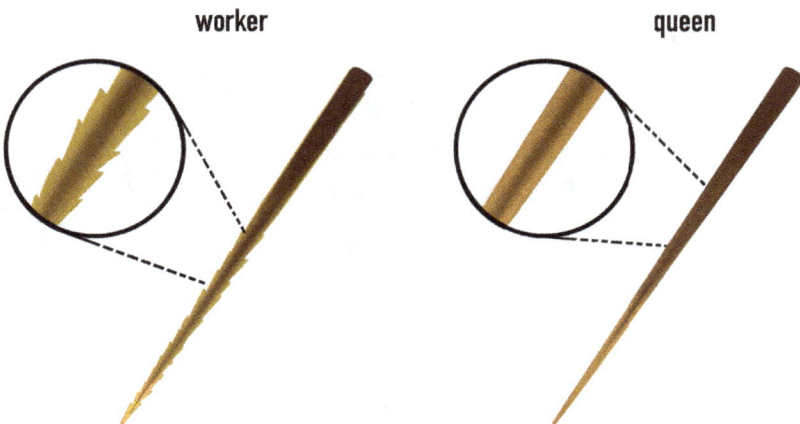

The difference between a worker bee's stinger and that of a queen bee

Worker bees have a barbed stinger. When we get stung, our skin is torn away during the bee's struggle to break free. This struggle is what generally kills the worker.

Bee sting venom has a protein that affects our skin cells and immune system, causing pain and swelling around the sting. If someone is allergic to bee stings, this venom will trigger a more serious immune system reaction.

When you hear beekeepers talking about "getting used to" stings, there is, in fact, a science behind this. According to a Yale School of Medicine study,[6] the key toxic component in bee venom can induce immunity and protect against future reactions to the toxin. So don't worry—although the first few stings might hurt, it only gets better from there.

Unlike worker bees, queen bees have a smooth stinger. It can be used more than once without causing harm to the queen. It's certainly rare, however, to be stung by a queen bee since she spends most of her life in the hive.

Did you know that unlike their female counterparts, drone bees are not equipped with a stinger?

Watch how to quickly and safely remove a honeybee stinger:
https://beekeep.blue/AwM

---

6   https://news.yale.edu/2013/10/24/bee-sting-venom-can-help-develop-immunity-bee-stings

Lifecycle

Although numbers vary depending on the type of bee you keep, expect an average of 10,000–60,000 bees in any one hive at the height of summer. Over the winter, numbers drop. Unlike bumblebee colonies, honeybee colonies can last through the winter as long as they have enough food, stay warm and don't suffer from disease or predators. To survive the cold, honeybees huddle together to form a winter cluster. During this time, the queen stops laying eggs, so there's no longer any brood to look after.

When spring comes around, up to 20,000 workers can still be left in a hive along with the queen. As the days slowly warm up and flowers are in bloom once again, honeybees emerge to forage for food, and the queen resumes her egg-laying.

Life Stages

Regardless of their role or duty, all bees go through four life stages: egg, larva, pupa, and adult. Reaching the adult stage depends on the type of bee: queens take 16 days to mature, worker bees take 21 days and drones take 24 days.

STAGE 1: EGG

Up to 3,000 eggs are laid daily in the hive, one per cell. The eggs are around the size of a grain of rice and stand upright until about day three, when they fall on their side. Millions of sperm are stored in the queen's body so she can fertilize her eggs when needed. Fertilized eggs are laid in smaller cells than those that aren't fertilized and become worker bees. Unfertilized eggs later become drones.

STAGE 2: LARVA

Four days after an egg is laid, a larva hatches, resembling a tiny worm with no legs, wings or antennae. If a larva is destined to be a queen, it will be fed royal jelly exclusively, which contains more honey and pollen than the bee bread mainly provided to workers and drones. Royal jelly allows potential queens to grow bigger than other female bees.

STAGE 3: PUPA

While larvae are kept in cells capped with wax, they spin a cocoon around themselves so they can pupate. The pupa eventually grows a head, eyes, thorax, abdomen, wings and legs, giving it the appearance of an adult bee.

STAGE 4: ADULT

Depending on the type of bee, 16 to 24 days after the initial laying of an egg, the adult bee chews its way through the wax and out of its cell to begin working.

Lifespan

Bees have a relatively short lifespan compared to humans and most other animals, but this doesn't detract from their incredibly intricate way of life. Depending on the type of honeybee, they have different expected lifespans:

- Queen bees tend to live between three to five years.
- On average, drones live for 55 days, although some beekeepers have observed drones surviving up to 90 days.
- Workers that hatch in the spring and summer have shorter, busier lives than those that hatch later in the year. Early workers will live for around six to seven weeks, during which they're hard at work.
- Later workers have less to do because there is no brood. Their main feat is surviving the cold until spring comes around, meaning they can live up to six months.

The honeybee lifecycle

Breeds

In the U.S., there are six main honeybee breeds, all of which live in a colony with a queen, make wax comb and produce honey. Understanding the differences between honeybee breeds will help in choosing the best variety to start your hives. Be aware that there can be some crossover between varieties, so it isn't always possible to keep your stock completely "pure."

Whether you're looking for a docile breed like the Italian honeybee or one more resistant to disease like the Carniolan variety, knowing about your breed's particularities and needs will allow you to make more informed choices and provide better care for them.

Italian or Ligurian Honeybee (*Apis mellifera ligustica*)

Italian honeybees are one of the most popular breeds to keep (and not just in Italy). As the name suggests, they're originally from the land that brought us Michelangelo and da Vinci, making their way to the U.S. in 1859.[7] Due to their color, Italian honeybees are attractive and easily identifiable with many desirable qualities.

Physical attributes: Light in color, to the point of a pure yellow in some colonies.

7 https://wisconsinpollinators.com/Bee/BA_HoneyBeeBreeds.aspx

Temperament: Gentle.

Characteristics: They're clean and resistant to disease. They tend to rob honey from weaker or neighboring hives.

Climate: Due to their Mediterranean heritage, they do well in most climates, except for tropical areas. Italian bees require more food in cooler northern latitudes to compensate for the heat loss due to loose clusters.

Brood rearing: They tend to raise their brood in late autumn and have extended brood-rearing periods.

Swarming: Moderate when honey stores are depleted.

As wonderful as they are, there are some drawbacks to keeping Italian honeybees:

- Compared to other varieties, they require more resources for longer because of their long brood cycles.
- Their interaction with other colonies due to their tendency to rob can increase the spread of pests and diseases despite their natural resistance to them.
- They're prone to hive drifting and may decide to leave their home colony to join another.

Carniolan Honeybee (*Apis mellifera carnica*)

As the second most popular bee variety in the U.S., Carniolan honeybees experience a massive build-up in the spring, making them perfect for beekeepers who want to increase their stocks before summer.

Physical attributes: They appear gray in color due to the tiny hairs covering their body. They're smaller compared to other European breeds.

Temperament: Docile and won't sting unless heavily provoked.

Characteristics: Their longer-than-average tongue (measuring up to 6.7 mm) helps pollinate crops like clover and gives them more sources of nutrition than other honeybee varieties. They also have a reasonably high resistance to certain diseases and parasites that can take out other breeds.

Climate: Originating from Central and Eastern Europe, they often struggle in the summer heat but have a high tolerance for the cold. They do well over winter, clustering tightly with a reasonable amount of food.

Brood rearing: Carniolan honeybees adjust their worker population to match the availability of nectar. Generating high numbers of workers during times of high nectar availability means they can store plenty of honey and pollen. Once nectar is no longer as abundant, brood production stops.

Swarming: They have a high tendency to swarm, possibly due to their rapid growth and comb production in the spring. Swarming can also occur later in the year because they don't need as much food as other breeds.

The downsides to keeping Carniolan honeybees:
- They don't produce as much wax and comb as other breeds.
- Their brood nest relies heavily on the availability of pollen.
- Their dark-colored queen can be difficult to find.

Russian Honeybee
(Apis mellifera from the Primorsky Krai region)

The Russian honeybee is originally a hybrid of Italian, Carniolan, Caucasian, German black and other bees from south-eastern

Russia. It was introduced to the U.S. in 1997 by the USDA after a rise in colony collapse due to parasites.[8] It's worth mentioning that Russian queens sourced from certified Russian Honeybee Breeders Association members are taken from a carefully protected, closed breeding population.

Physical attributes: Ranges from yellow to black in color.

Temperament: Tends to be more aggressive than other bees.

Characteristics: Natural tolerance to Varroa and tracheal mites.

Climate: They fare well in cooler temperatures. While they winter in smaller populations, Russian honeybees will often overtake the hive population of neighboring Italian honeybees in the spring when pollen and nectar become available again.

Brood rearing: There is extensive brood production with the availability of pollen, resulting in a strong spring buildup.

Swarming: Moderate. Populations can explode to the point of overcrowding, so these bees will swarm if they don't have enough room. To avoid swarming, ensure the queen has all the space she needs to lay eggs and store honey.

Other behavioral peculiarities of the Russian honeybee:

8   https://wisconsinpollinators.com/Bee/BA_HoneyBeeBreeds.aspx

- Whereas most breeds only create queen cells during swarming or queen replacement, Russian honeybee colonies usually maintain queen cells at all times.
- They are strong honey producers.
- When Russian bees are around other breeds, they tend to crossbreed, diluting the stock and increasing the risk of natural pests.

Buckfast Honeybee (Apis mellifera)

Buckfast honeybees are a hybrid variety first created at Buckfast Abbey, in the United Kingdom. In the early 20th century, bee populations were under serious threat from tracheal mites.[9] Karl Kehrle, also known as Brother Adam, was the monk in charge of beekeeping at the abbey. To protect the bees from tracheal mites, he crossbred the strongest surviving colonies, giving us the Buckfast honeybee—a popular choice for apiarists living in the UK and similar climates, but less common in the U.S.

Physical attributes: Their color resembles the older, darker Italian honeybees. There can be a variation in their appearance due to crossbreeding.

9 https://wisconsinpollinators.com/Bee/BA_HoneyBeeBreeds.aspx

Temperament: Buckfast honeybees are among the gentlest bees around, so a smoker is rarely needed when working with them.

Characteristics: They have a high resistance to some natural parasites, are skilled at foraging and produce plenty of honey while consuming less.

Climate: These bees can survive the winter well.

Brood rearing: They have highly fertile queens. This breed tends to increase their population rapidly in the spring and maintains their numbers throughout the summer.

Swarming: Rare.

Some potential drawbacks of keeping Buckfast honeybees:

- They produce less propolis than other varieties.
- They can suffer from inbreeding, reducing their natural resilience to pests and other positive attributes.

Caucasian Honeybee (*Apis mellifera caucasica*)

Caucasian honeybees were once highly popular in the U.S., but their popularity has decreased among beekeepers looking to maximize honey production. However, their longer-than-average proboscis and gentle nature still make them good company among commercial pollinators.

Physical attributes: This is a hairy breed, ranging from silver-gray to dark brown in color.

Temperament: Gentle.

Characteristics: You'll often find they cover everything in a sticky layer of propolis, as they use a lot of it.

Climate: Originating from the Caucasus region between the Black and Caspian Sea (an area with wildly differing climates), they can easily adapt to changing conditions.

Brood rearing: They take longer to build their numbers in the spring than Italian bees do.

Swarming: Rare.

The disadvantages to keeping Caucasian honeybees:

- They're prone to nosemosis.
- Their high use of propolis makes it harder to manage the hive.
- They tend to rob honey from other colonies.
- If you live in an area with a high nectar flow in spring, you're better off going for a different variety, as Caucasian honeybee colonies don't reach full strength until later in the year.

European Dark Honeybee (Apis mellifera mellifera)

Also known as the German black bee, this breed came to the U.S. from Northern Eurasia during the colonial era.

Physical attributes: European dark honeybees have a stocky body and minimal brown abdominal hair. Their wings are dark in appearance, and they seem almost black when viewed from a distance. They're recognized by the lighter, yellow bands on the sides of their abdomen.

Temperament: They are defensive and have a reputation for stinging for no apparent reason! Certain wilder hybrids of this variety can be more on the aggressive side.

Characteristics: Over time, a range of subspecies developed because of the breed's resilience. Characteristics can vary wildly.

Climate: They're well-suited to the climate in the UK (cold and damp). This breed has a good survival rate throughout long, cold winters.

Brood rearing: They grow more slowly than other breeds. Their population is built up later in the year, so they're unable to take advantage of an early spring nectar flow.

Swarming: Moderate.

Other drawbacks of keeping European dark honeybees:
- Their comb tends to be runny.
- They generally produce less honey than other honeybee breeds.

Set yourself up for success by choosing the best type of honeybee for your climate and style of beekeeping. Remember that it's always helpful to further research your particular geographical location and climate.

	ITALIAN	RUSSIAN	CARNIOLAN	BUCKFAST	CAUCASIAN	EUROPEAN DARK
Temperament	Gentle	Slightly more aggressive	Gentle	Gentle	Gentle	Defensive/aggressive
Over-wintering	Good	Very good	Good	Very good	Good	Very good
Swarming	Moderate	Moderate	High	Rare	Rare	Moderate
Brood rearing	Late autumn over extended periods	Generally low numbers	Responsive according to nectar flow	Rapid increase in the spring	Slower than Italian bee	Slower than others
Resistance to pests & disease	Low to moderate	Moderate	Moderate	Moderate	Moderate to high	Moderate to high

"A co-worker called one day and asked if I was interested in taking his hives. I said yes but had no clue about beekeeping at the time. That very day, I had the hives at my house and started watching YouTube to figure out what to do with them. I came across Jason's channel, and here I am 6 years later..."

—Plan Bee Apiaries

CHAPTER 3

Hive Essentials

Knowing about different kinds of hives, hive parts and the function they each serve is the best way to troubleshoot any issues that come up along the way. The more we know about something, the more we can do something about it—whether we're talking beehives or whatever it may be!

If you're a skilled woodworker and want to invest in building a hive from scratch, there are plenty of resources out there for you. My only concern with DIY hives is ensuring appropriate bee space—if the specs aren't exact, your frames will likely become stuck together, creating more problems and work. As a beginner beekeeper, I suggest purchasing your first hive. The following information can help you choose the right one.

In this chapter, I'll also discuss the incredible synergy that takes place within a hive between the queen, worker bees and drones. With each bee assigned a specific duty, what results from this intricate network is a unified whole greater than the sum of its parts.

Types of Hives

The two most popular choices for beekeepers are the Langstroth and the top bar hives. All hives have pros and cons, which means it isn't one size fits all. While some beekeepers might tell you their type of hive is the only one worth using, the best kind is the one that's best for *you* and your needs.

The hive you choose will be a home for your bees for years to come, so it's important to consider the particularities of your climate. You'll also be working with your hive regularly, so think about how much time you'll have to devote to maintenance. Choose wisely!

Langstroth Hives ($150–$300)

Invented by Reverend L.L. Langstroth back in 1852, these hives are the most common design used in North America today. Langstroth was a beekeeper based in Ohio who wanted a hive that would allow him to view his colonies' brood and honey production with minimal disruption. While his original design has since evolved, it's still based on a modular, expandable beehive that's easy to access.

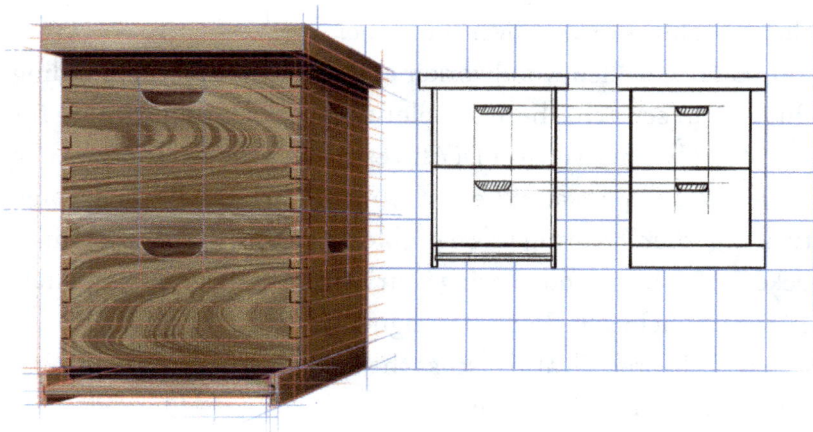

Langstroth hives are the most common design in North America

Langstroth hives feature removable hanging vertical frames where bees can build their comb. The gaps between each frame and the box's perimeter are precisely measured to give bees the perfect amount of space. By leaving ¼ to ⅜ of an inch between frames, the Langstroth hive ensures bees won't fill these spaces with propolis or join them with comb, facilitating hive management.

The notion of bee space was identified by Langstroth and is used in all good hives, regardless of design.

There are two sizes of Langstroth hives: eight-frame and ten-frame. Langstroth hives can be expanded by putting new boxes on top of the existing ones—these upper-story hive boxes are called supers and are placed over the brood chamber so excess honey can be stored.

Due to their standard sizes, it's easy to source replacement parts for this kind of hive from different manufacturers. Resources and accessories are also widely available. If you want a large apiary or are mainly focused on honey production, Langstroth hives are a good choice. However, when their boxes are full of honey, they can weigh anywhere from 30–100 lbs, making them hard to lift and move around. If weight is a deciding factor for you, consider the eight-frame setup over the ten-frame.

Top Bar Hives ($200–$400)

Top bar hives are designed differently from Langstroth hives. Standing at a convenient height, they're the most comfortable hive to work with, as there is no heavy lifting required.

Top bar hives are also wider than other hives and have an extended roof to protect the bees and their honeycomb. Beneath the roof are 24 wooden bars, which give the hive its name. Each of these bars has a starter strip used to build comb. Although this

The top bar design places less pressure on a beekeeper's back

design forces bees to move across the bars instead of up and down (as they prefer), they eventually adapt and get used to moving from side to side.

Since there is no foundation, the comb in a top bar hive is delicate, requiring more care and maintenance than others. When it comes to honey extraction, frames must be crushed and the honey must then be strained. On the bright side, they don't need any special equipment, which makes them one of the more economical choices.

Top bar hives are very similar to horizontal Langstroth hives. If you already own a standard Langstroth, consider the horizontal Langstroth over the top bar hive, since its frames can be interchanged if needed.

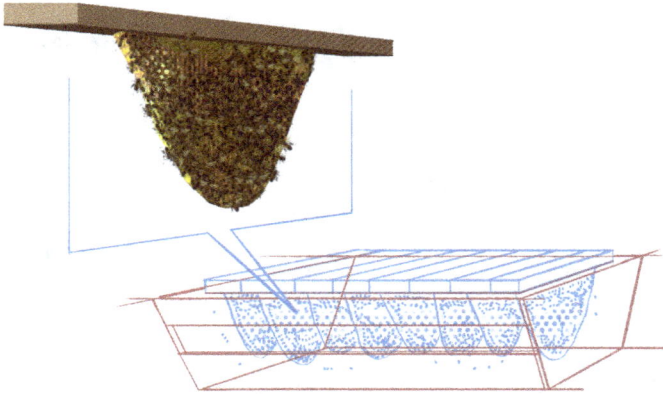

Top bar hives use bars equipped with a starter strip for building comb

Apimaye Hives ($300–$500)

The Apimaye is the Cadillac of Langstroth hives, equipped with upgraded features, such as a pollen trap, an integrated hive feeder and a ventilation system. I personally love my Apimaye hives.

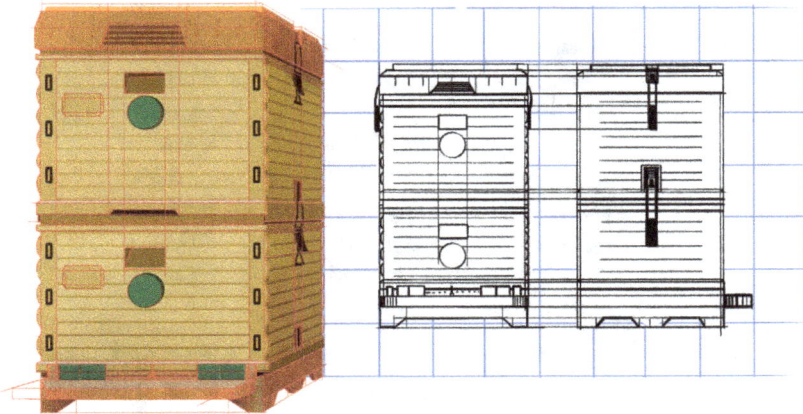

Additional features make Apimaye the Cadillac of Langstroth hives

Given their double-layer insulation, Apimaye hives are protected against extreme temperatures. A special material in the hive's top cover and side walls provides thermal insulation. Since Apimaye hives aren't made of wood, the good thing is they don't absorb moisture, which lowers the chances of fungus growth.

As much as I'm a fan of Apimaye hives, I suggest skipping the extras and keeping things simple when first starting. Standard Langstroth hives have my vote.

	Langstroth	Top bar	Apimaye
Cost	$150–$300	$200–$400	$300–$500
Pros	• Standardized design • Expandable • Efficient honey extraction • Incorporates bee space	• Allows for more natural bee behavior • Comfort and accessibility • Less equipment required • Cost effective	• Insulation properties • Convenient features • Resistant to moisture, pests and rot
Cons	• Heavy lifting when full • Additional equipment required • Preventive maintenance required for weather protection	• Delicate comb • Additional maintenance required	• High cost • Lower compability and availability compared to traditional designs

Hive Parts

The standard Langstroth hive will be the reference model for the remainder of the book. If you decide to go with a different type of hive, most of the book's contents will still be useful, but the immediate section that follows less so.

Let me show you the
parts of a Langstroth hive:
https://beekeep.blue/OtL

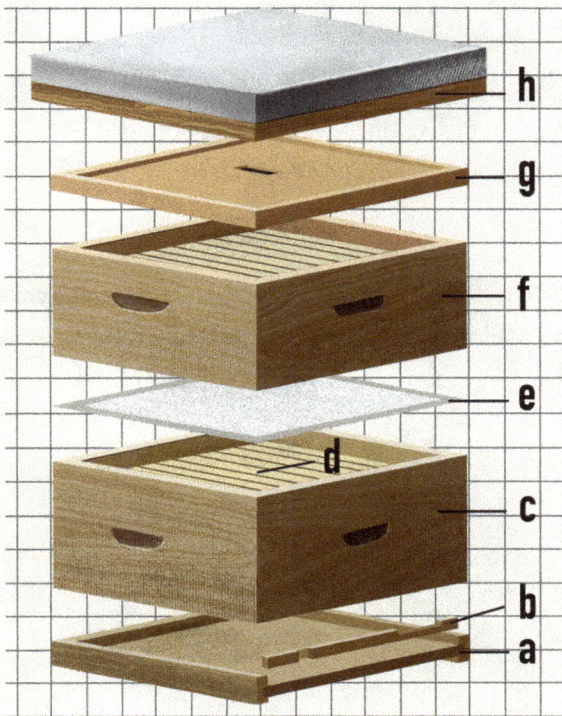

The parts of a Langstroth hive. From the base moving upwards

a) Bottom board

The bottom board is at the very base of the hive and protects bees from small hive beetles, Varroa mites, skunks, raccoons, etc. There are two types: solid and screened. Both serve various purposes and can be switched out according to need.

- A **solid bottom board** is made of wood. To protect it, it's a good idea to use a hive stand to keep it elevated off the ground. This kind of board is part of the original Langstroth design and is often used by beekeepers trying to mimic a bee's natural habitat. It helps keep heat in over the winter, encouraging an early start to brood production in the spring. Moreover, solid bottom boards support better communication within a hive, since pheromones are more likely to be contained within the hive.

Bees use multiple pheromones to communicate with each other, such as when there's a threat to the hive or the colony requires a new queen.

- The floor of a **screened bottom board** is made of woven metal. This creates ventilation, resulting in lower moisture levels inside the hive—particularly helpful in warmer climates. The airflow

A screened bottom board promotes ventilation in a hive

also reduces the chances of overheating. When bees overheat, they beard on the face of the colony, so having a screened bottom board reduces this likelihood.

Many northern beekeepers, including myself, use screened bottom boards year-round. I learned that my bees had no problem dealing with the cold air. They have survived winter just fine, so I've simply stuck with what works. Although brood production may be slightly slower come spring due to the airflow, it's nothing to worry about.

If you're in a desert-like area, you may want to consult a local beekeeper for advice because a screened bottom board may leave your bees too dry, affecting their overall brood production.

STICKY BOARD

A sticky board can be placed under the wire base of a screened bottom board. It is usually made from a thin piece of wood or corrugated plastic covered in something sticky (like coconut oil or vegetable shortening) and can be used to catch and count Varroa mites.

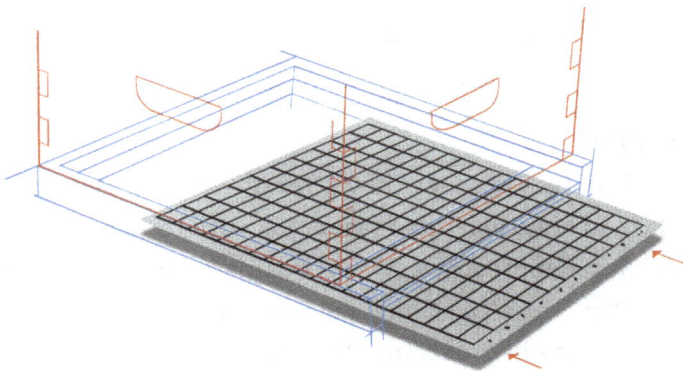

A sticky board is used with a screened bottom board for catching and counting mites

b) Entrance reducer

This defends your colony from intruders like wasps, hornets and other bees. Place the **entrance reducer** at the front of your hive between the bottom board and the lip of the bottom brood box. Adjust the size of the opening as needed.

Bees are weak and vulnerable at the start of the year, so you should set the entrance reducer to the smallest opening. As the year progresses and your colony becomes stronger, you can gradually enlarge the entrance before removing this piece entirely.

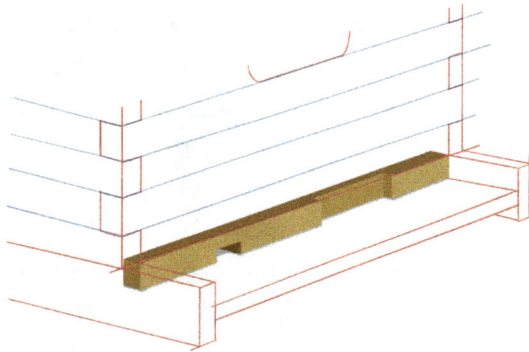

*By restricting the size of a hive's opening,
an entrance reducer protects against intruders*

MOUSE GUARD

Mice love beehives because they offer plenty of free food and a warm place to stay. For this reason, it's wise to install a **mouse guard** as winter approaches. Mouse guards are made of metal, so mice can't chew through them. Like an entrance reducer, a mouse guard fits over the hive entrance for protection.

For an easy DIY mouse guard, staple rabbit wire over your hive entrance, and that should do the trick!

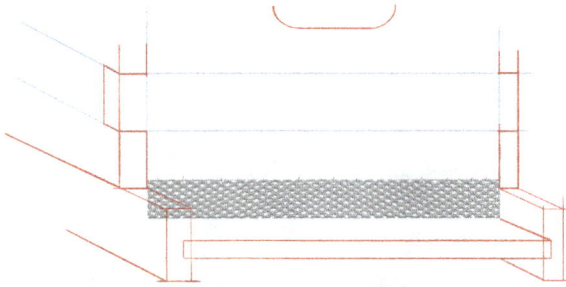

*Like an entrance reducer, a mouse guard fits over
the hive entrance to protect your hive*

c) Brood chambers

The brood chamber consists of a deep box filled with frames and
sits on the bottom board. In an active hive setup, there are usually
two deep brood chambers. This part of the hive is where the queen
bee lives and lays her eggs. It's also where larvae transform into
pupae and later mature into adult bees. In the brood chamber,
worker bees are busy carrying out their duties.

d) Frames & foundations

In Langstroth hives, the frames keep comb organized in both the
brood chambers and honey supers. Each frame has two sidebars,
and a top and bottom rail, forming a rectangle that's held together
with wood glue and/or nails (preferably both).

Frames also consist of a flat board covered in hexagon cells, also
known as a **foundation**. Made from either plastic or wax, these
cells serve as a guide for bees to build their comb straight down
the length of the frame. Without foundation, bees tend to fill the
spaces between frames with honeycomb, making removal more
difficult during inspection.

Frames can also be foundationless. At the top of foundation-less frames, a starter strip made of beeswax or plastic serves as a starting point and guide for bees to build their comb.

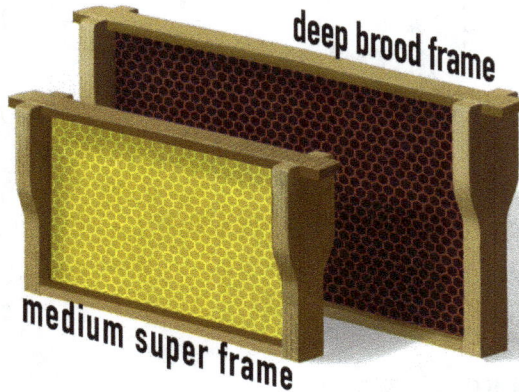

Frames keep the comb organized

e) Queen excluder

The purest honey comes from frames that have never contained brood. When the queen bee moves into the hive's upper boxes and lays eggs next to the honey, you'll have to wait for the eggs to hatch before harvesting. Even then, leftover cocoons will usually turn dark brown, leaving remnants of the eggs and other hive debris behind.

To ensure that your honey is as pure as can be, keep the queen out by placing a queen excluder between the top brood box and honey super. A queen excluder is a plastic or metal screen with gaps large enough for worker bees to pass through but too small for the queen.

A queen excluder can also be used to help the colony settle into its new home when first installing your bees. Place it between the brood box and the bottom board. If your hive has a top entrance, close this off, allowing the worker bees to still leave the hive from the bottom of the hive while containing the queen inside the brood box. Once the queen starts laying eggs, you'll know the colony is there to stay. The queen excluder can then be removed.

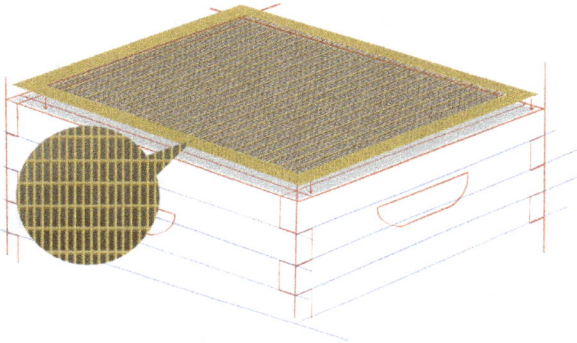

A queen excluder has multiple purposes,
including keeping honey pure

f) Supers

This is where bees store their honey. Supers usually consist of shallow, medium-sized boxes that can get quite heavy once filled. Multiple supers can be added to a beehive as needed. Within the supers are **frames**.

Before the invention of the Langstroth hive, beekeepers had to destroy the comb in a hive to get to the honey, often killing the colony in the process.

g/h) Inner & outer covers

The top lid of the beehive is called the outer cover and protects the hive from the elements. The inner cover is placed directly below the outer cover. It prevents the super from sticking to the roof. Inner covers are usually made of wood and have a hole in the middle, helping to regulate the hive's temperature.

Hierarchy & Duties

Every hive has one queen with an abundance of female worker bees and a population of about 10–15% male drones. Keep in mind that these numbers are during the growing season, as drones are later booted from the hive in winter.

When I first started beekeeping, I found it fascinating to learn that every bee has a job—a way through which they each contribute to their colony. Although there is a hierarchy, every role is equally important. The fact that no bee can survive without the others qualifies a bee colony as a superorganism.

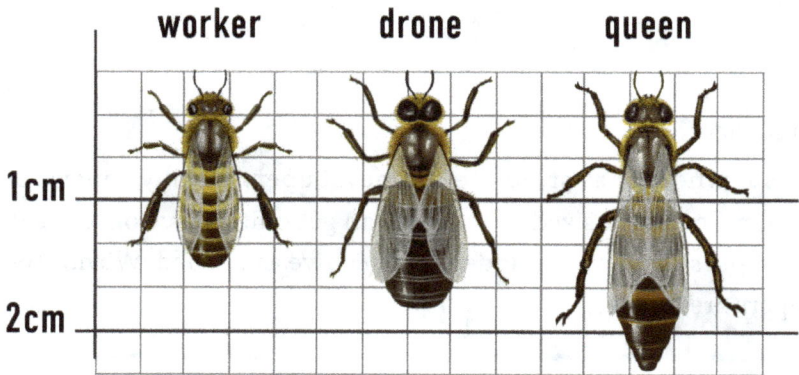

A worker, drone and queen bee have slightly different anatomies to suit their specific roles in a hive

The Queen Bee

As you already know, the queen bee is the center of every hive. A colony is born with the mating of a queen honeybee. She does this only once when she leaves the hive to go on a mating flight, procreating in the air with drone bees as many as 24 times. The queen then returns to the hive to start laying her eggs.

She spends her life laying and fertilizing her eggs to keep the hive going, constantly giving off pheromones to ensure that female workers remain sterile. These pheromones also let the rest of the bees know that the queen is alive and well.

Because it can sometimes be difficult to pick out the queen from other bees, beekeepers often mark the queen's thorax with bee paint so she can be identified more easily.

Without a queen, a hive loses direction: foraging activities decrease, and when worker bees do go out, they tend to bring back less nectar and pollen. When a queen dies or her egg production slows, worker bees will raise new queen prospects.(More information can be found in Chapter 6.)

Male Drones

Drones have only one—albeit crucial—job: to mate with the queen. Once accomplished, that is the end for them. Adding to this almost tragic existence, even while alive, male drones are unable to feed themselves as they're not equipped with the body parts to be able to forage. They're also unable to defend themselves since they don't have a stinger, and when food is sparse or winter hits, drones are left to starve after being pushed out of the hive by worker bees.

In response to Beyoncé's question, "Who run the world?", there's no denying that in the world of bees, it's also the "GIRLS."

Female Workers

Female bees are the productive ones in a colony, which is why they're called "workers." They outnumber male drones 100:1 and do all they can to maintain the hive except reproduce (sometimes, female workers will lay eggs, but we'll get to that a little later).

A female's responsibilities are extensive and change over their lifespan. Responsibilities are divided into **house** and **forager** roles.

HOUSE ROLES

- **Honeycomb builders** make beeswax and build honeycomb. Hundreds of bees can work on the same comb area and must eat large quantities of honey and nectar to make enough wax. It takes around 12 hours to make eight wax scales (approx. 1,000 scales = 1 g of wax).
- **Nurse bees** feed and care for larvae, checking in with them over a thousand times daily. When a parasite infects a hive, nurse bees will eat honey with a high antibiotic content and feed it to the rest of the colony for protection.
- **Queen attendants** groom and feed the queen. Every queen has her ladies-in-waiting.
- **Groomers** clean the other bees, rapidly removing dust, stray hairs and debris.
- **Cleaners** clean used cells or clear away debris from the hive. If the queen isn't happy with their work, they're asked to do it again.
- **Undertakers** take care of dead bees. While most bees die outside the hive, those that don't are removed. Once they've dried out, undertaker bees take them far away, so the bodies don't attract pests or disease.
- **Honeycomb cappers** secrete beeswax to seal pupae cells for protection, which they also do for cells filled with ripened honey.
- **Pollen packers** collect the pollen brought back by foragers and place it into cells to be eaten later.
- **Nectar ripeners** transform nectar into raw honey by placing it into cells and fanning it to evaporate excess water.

- **Bees that repair and fix** use propolis to mend cracks in the hive and cover up any foreign objects too big to be removed.
- **Temperature controllers** maintain the temperature and humidity of the hive. If the temperature gets too high, water is fetched and spread on the backs of bees that fan their wings, creating airflow and lowering the temperature with the evaporated water.

FORAGER ROLES
- **Propolis collectors** retrieve the resinous substance from trees and bring it back to the hive in their pollen baskets.
- **Water fetchers** make up 1% of the hive's population and are devoted to this vital task. Water cools the hive and is also used by nurse bees to dilute raw honey, which is then fed to larvae.
- **Guards** check the scent of every bee coming into the hive, allowing only colony members through. They also inspect the hive for cracks, guarding against robber bees or other intruders. When the hive is breached, guards use their stingers to defend the colony and give off pheromones to warn of approaching danger.
- **Nectar and pollen gatherers** leave at sunrise, making about ten foraging trips daily. Every trip, the bees visit up to 100 flowers, filling their honey sacks with nectar and their pollen baskets with pollen. These goodies are then transported back to the hive.

PART II
Building Momentum

"I started beekeeping in 2014 when my son and his neighbor invited me to join them. The main reason was that I had been gardening for about 50 years and wasn't seeing a lot of honeybees. Beekeeping was a new hobby during my retirement, and we have enjoyed the sweet rewards ever since. Honeybees and the structure of their work are something to behold."

—Dennis Michael

CHAPTER 4
Setting Up Shop

When I first started beekeeping, I wanted to ensure I had everything I needed and that my yard had everything my bees needed to thrive. I also spent a lot of time researching where to get my bees. The options were overwhelming: Should I start with packaged bees? A nuc or established colony? What's the difference between them all? Do they come with queens?? So many questions…

Well, folks, now that the foundation has been laid for how bees operate and how hives work, let's get you set up and look at what you need to start taking action. As you go through this chapter, you'll learn where to place your hive and how to create a flourishing environment for your bees. I'll also outline the essential beekeeping equipment needed. After that, I'll share how to source your bees and what to do once you finally receive them.

If you're in the U.S., check with your state's Department of Agriculture to see if registering beehives is necessary. In Ohio, it costs $5 per bee yard regardless of the number of hives.

There's No Place Like Home

Bees are highly intelligent creatures. They will learn the location of their hive, so when choosing a spot in your yard, be aware of the "three-foot or three-mile" rule. According to this rule, a hive can be moved up to a radius of three feet before your bees become confused—any further than that and you'll need to relocate the hive three miles away from the site for an entire week before moving it to its new location.

While this rule has certain exceptions, it's best to stick to these guidelines as a beginner. Choose your hive location wisely because changing your mind isn't as simple as rearranging the living room furniture!

Watch this detailed video explanation of the "three-foot or three-mile" rule:
https://beekeep.blue/DyH

Of course, where you ultimately place your hive will depend on how much space is available, the proximity of your land to that of your neighbors, the landscape, local regulations, etc. You may also want to consider the following:

a. Finding a dry, flat, low-traffic area where your bees are less likely to be disturbed.

b. Placing your hive somewhere that causes minimal disruption to your surroundings while keeping easy access.

c. Fencing off the area if there's a possibility that children may go near the hive.

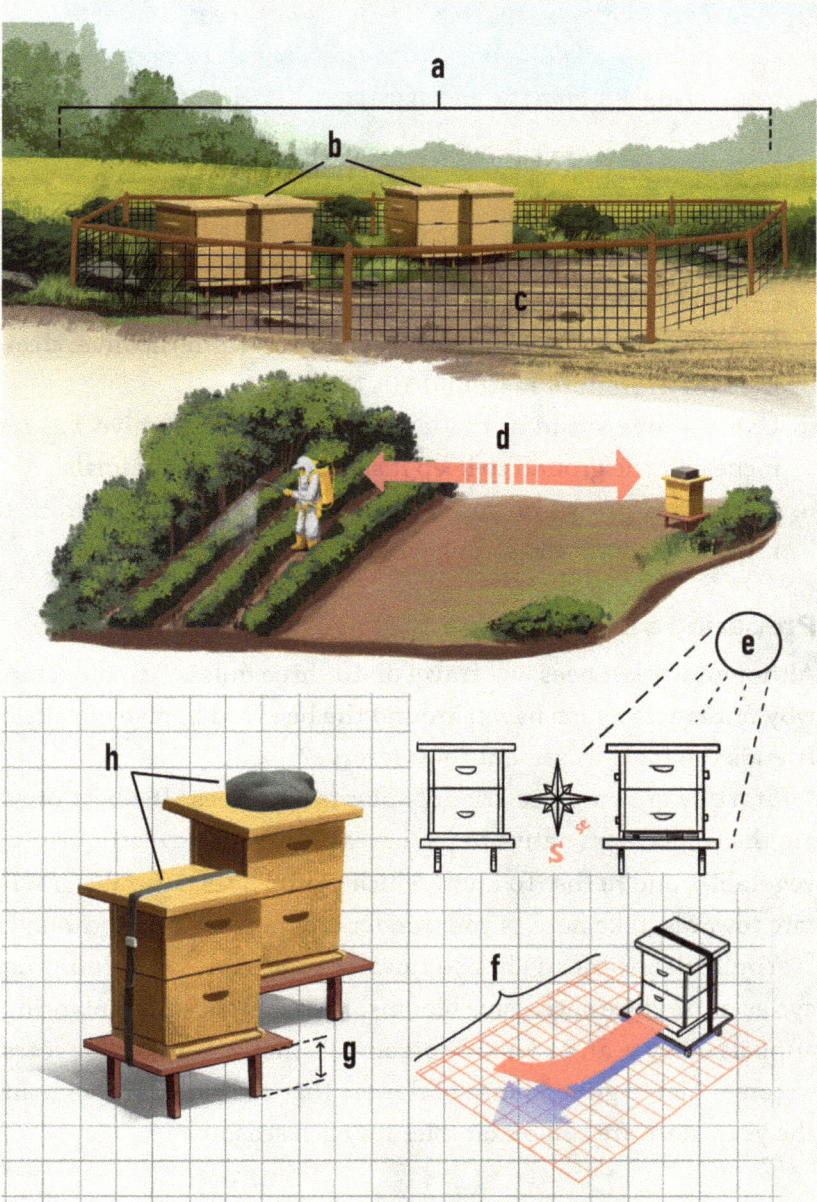

Tips on where and how to place your hive

Having a six-foot privacy fence will force bees to fly up instead of straight out—especially handy with neighbors nearby since your bees will fly above their land instead of across.

d. Keeping your hive as far as possible from areas where pesticides are being used.

e. Orienting your hive south or southeast to get six to eight hours of direct sun. More sunlight promotes your bees' productivity.

f. Giving your bees a clear flight path to and from the hive: clear the area in front of it around 10–15 feet.

g. Using a hive stand or cinder blocks to get your hive 12–18 inches off the ground to keep it safe from smaller animals.

h. Placing a large rock on your hive or strapping it down to protect it from raccoons and strong winds.

Preparing a Bee Paradise

Although worker bees will travel up to three miles to find nectar, why not create a lush haven around the bee yard to give our little friends the royal treatment they deserve?

Start by growing bee-friendly plants and trees. The best ones are the wildflowers native to your area, as well as flowering fruits, vegetables and herbs. To know which plants bees naturally gravitate towards, take note as you wander around your neighborhood.

You'll also want to choose flowers that have long blooming cycles or produce successive blooms. If you're not up for planting annuals every year, pick hardy perennials that will flower for years to come. Aim to grow a range of flowers that will bloom throughout the year. Here are some examples for each season:

LATE WINTER - EARLY SPRING

- borage
- California lilac
- California wildflowers
- manzanita
- native blueberry
- redbud
- red maple
- rosemary
- titi

LATE SPRING - SUMMER

- apple
- bee balm
- blackberry
- black locust
- buckwheat
- bugleweed
- cabbage palmetto
- calamint
- canola
- chaste tree
- crimson clover
- Dutch clover
- gallberry
- heavenly bamboo
- henbit
- honeysuckle
- jessamine
- lavender
- multibranched sunflowers
- oregano
- persimmon
- privet
- red clover
- skip laurel
- sourwood
- tallow tree
- tulip poplar
- tupelo

LATE SUMMER - EARLY WINTER

- American germander
- aster
- Brazilian pepper
- Brazilian verbena
- bush clovers
- butterfly bush
- buttonbush
- coyote bush
- crape myrtle
- cup rosinweed
- glossy abelia
- goldenrod
- kenaf
- Maximilian sunflower
- meadow beauty
- Mexican marigold
- Mexican sage
- Mexican sunflower
- milkweed
- mountain mint
- New England aster
- partridge pea
- pineapple sage
- purple coneflower
- redroot
- stonecrop
- summer phlox
- sweet pepperbush
- thoroughwort
- wood mint

For more recommendations, your local plant nursery or arboretum is also a great resource to consult.

Another way to indulge your bees is by grouping plants. This arrangement makes it easier for them to find the ones they like. A good guide is to have at least one square yard of each plant. Keep it organic—avoid pesticides and herbicides. Weeds are your friends from now on. If they flower, bees will love them. Dandelion, clover, goldenrod and milkweed are all great!

Finally, make sure your bees have easy access to water. Keep a shallow source available by leaving saucers of water out every day. And, if you *really* want to spoil your bees, install a water garden with floating plants and a few rocks for your bees to land on. They'll be in absolute paradise!

Your bees will love having a small water garden around

Protective Clothing, Supplies & Tools

As a beginner beekeeper, I didn't have the money to buy all the protective gear I needed. My suit consisted of two sweaters and a light tan button-down shirt. I'd pull my socks over my pants and lace my boots up tightly. My veil, which I inherited from my beekeeping grandfather, looked every bit of its 25 years. The hive boxes that were also passed on to me needed repair and replacement.

These days, when I'm working with my bees on a hot summer day, I can be found in shorts, a t-shirt and nothing but a face cover. My veil is the one thing I never do without—stings to the face or head are incredibly painful and can be dangerous. I might get stung elsewhere, but that doesn't bother me much anymore.

Your beekeeping days probably won't start like this, but one day, you may reach this same level of comfort! Below, I share the essential pieces of clothing and equipment to acquire as you set out on your beekeeping journey.

**Watch a video walkthrough of
all the equipment I use:**
https://beekeep.blue/Vuj

Bee Suit ($95–$260)

Covering you from head to toe, a **bee suit** is made of heavy fabric and comes with elasticated cuffs to make you sting-proof. It usually includes a hat and veil. While getting stung isn't as common as you might think, having a suit provides peace of mind so you can be more relaxed around your bees.

Beekeeping suits are traditionally white since the common lore is that for bees, wearing dark colors can make you look like a predator such as a bear. Quite honestly, I've worn all kinds of colors, and I've never noticed a difference in my bees' behavior.

Ventilated suits also keep you cool and protected on those 90°F (32°C) days. If you can't afford a bee suit right now, do what I did—improvise! Layer up and use duct tape to seal any openings. Just make sure you at least have a veil.

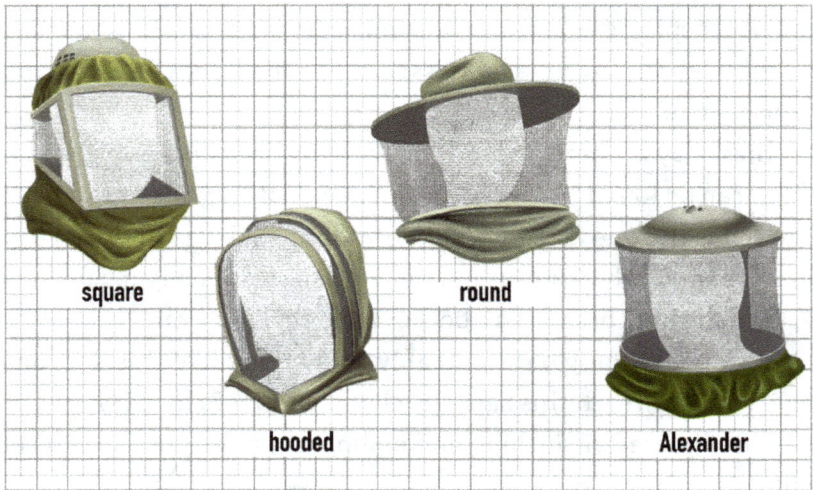

Types of veils

Veil ($20–$50)

Your choice of veil will come down to personal preference. Here are a few options:

- **Round veils** provide a full range of vision with 360° mesh (and occasionally a seam at the back). They work with a helmet or hat that may need to be bought separately.
- **Square** or **folding veils** go over a helmet or hat. These have several seams to support the structure and keep the veil away from the face to protect your vision. They can be folded flat.
- **Fencing** or **hooded veils** zip onto a bee suit or jacket. They have mesh on the front and some on the sides. There's no mesh on the back, though, so there's less airflow than with other veils. Some of these veils are also available as a pullover, combining the veil with a shoulder covering for extra protection.
- **Alexander veils** are round with a cloth top and elastic headband. You don't need to wear a hat or helmet with this type of veil, although many people like to wear a cap underneath.

Gloves ($15–$30)

You'll need gloves that are thick but flexible so you can perform the more delicate beekeeping jobs. Leather gloves work fine for this. Most beekeeping gloves reach up to the elbow for extra arm protection.

Boots ($40–$50)

Boots should have a thick sole and fit snugly over your bee suit so bees can't get in. Alternatively, you might want to pull your pant legs over the boots and seal with duct tape. Using ordinary rain-boots is fine, although you can buy specialist beekeeping boots if preferred.

Bees tend not to sting feet and ankles, but if you drop a frame full of bees on your feet, who knows what can happen! Something I don't recommend is wearing sandals when working with your bees.

Hive Tool ($10–$20)

Versatile with an unimaginative name, a hive tool is used for everything from separating hive boxes to lifting frames without damaging the wax, in addition to removing wax and propolis from the hive. Although you can make do with other tools lying around for these jobs, a hive tool is inexpensive and worth the buy.

Smoker ($35–$60)

This, of course, is used to puff smoke around your bees, masking their pheromones so they stay calm while you work around them. To lower your chances of getting stung, do not start a hive inspection without a good smoker on hand.

Bee Brush *(under $10)*

A bee brush helps move your bees around your hive without hurting or upsetting them. Use this tool to brush bees off a frame during an inspection, nudge them into a box as they swarm or get them off your clothes after an inspection.

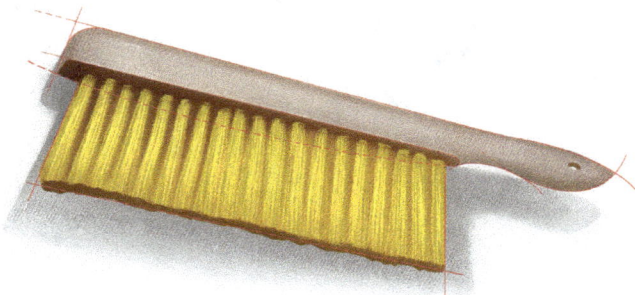

Note that not all bee brushes are created equal. Some are made with horsehair and do more damage than good: bees can get tangled in the hair, which they don't like, causing them to sting! Instead, invest in a good, soft plastic bristled brush (usually yellow). Your bees will love you for it.

Watch this for more info on smokers,
hive tools and bee brushes:
https://beekeep.blue/Piq

Feeder ($10–$30)

While bees are equipped to feed themselves, beekeepers should supplement their bees' diet when first installing them into a hive or during winter. Having a feeder when you first start beekeeping is essential. Although there are several types of feeders, some are better than others. The better ones are fully contained inside the hive, which reduces the risk of hive robbing.

See how different kinds
of feeders work:
https://beekeep.blue/koJ

Capping Scratcher or Uncapping Tool ($10–$40)

Use this narrow brush with sharp wire bristles to scratch open wax cells to retrieve honey or uncap single cells to check on larvae and inspect for signs of disease. While a regular sharp knife can also be used, a capping scratcher or uncapping tool makes the job much easier.

Sourcing Your Bees

Assuming you now have all you need to welcome your bees home, let's talk sourcing. While buying your bees is a fairly simple process, some planning is required. First things first, you should decide which breed of honeybee you want. Refer to Chapter 2 for more information and consult your local beekeeping organization.

Next, it's time to research suppliers. Look for reputation over price—cheaper isn't always worth the saving. I suggest you do your best to source them locally or from a region with a similar climate. Again, your local beekeeping organization or county extension agent can provide local supplier recommendations.

When ordering bees, it's also wise to talk with your postal carrier beforehand and work out a plan for their arrival. My postal carrier has me pick my bees up at the post office. Upon delivery, act quickly, since keeping them for too long at the postal building can lead to your bees overheating.

Watch this video to learn more about sourcing bees:
https://beekeep.blue/9po

Packaged Bees

The most common way to purchase bees is in packaged form. When you go this route, you'll receive a queen, along with roughly 11,000 other bees (the equivalent of about three pounds). The package also includes a feeder with sugar syrup.

Packaged bees come with roughly 11,000 bees

Packaged bees are bred primarily for sale. Most breeders in the U.S. are based in the southern states but will ship nationwide. Although purchasing locally is ideal, you may not have access to local packaged bees unless you're in the south. If you're based in a different region and still decide to buy packaged bees, it's not a big problem. I myself have had my share of successes with packaged bees that weren't sourced from my region.

For an extra fee, you can request that the supplier mark the queen so you can identify her more easily. The color used to mark the queen will reflect the year she was born—years ending in: 1 or 6 – white; 2 or 7 – yellow; 3 or 8 – red; 4 or 9 – green; 5 or 0 – blue

Nucleus Colonies (or Nucs)

Nucleus colonies include worker and drone bees, an active queen, several frames of brood at all stages, as well as a frame filled with nectar and pollen. When I mention "local" bees, they would come in the form of a nuc for most folks, which would consist of a thriving split from a local bee colony. These colonies can be sourced from private companies or local beekeepers.

With nucs, you get a head start with the establishment of your colony, which is why they can cost up to twice the amount of packaged bees. Not only do you get an actively laying queen that continues to lay during transportation, but your hive also benefits from bees of varying ages that already have the hang of working together—now there's a lot of bang for your buck!

While packaged bees must spend the first weeks in their new home drawing out comb for brood, nucs can start foraging and making honey immediately. There is a risk, however, that the honeycomb can bring pests and diseases from the donor hive to yours. It's always good practice to ask the supplier about state inspections and what treatments have been done to keep the bees healthy.

For more on buying a nuc vs. packaged bees, watch this:
https://beekeep.blue/laJ

Established Colonies

Buying an established colony is another way to hit the ground running, as it includes your hive and frames, along with all your bees. As a fair warning, established colonies tend to be more aggressive in defending their hives. Starting with more bees also makes it trickier for inspection, and it's hard to track how old the queen is. If the queen dies (knock on wood), you risk losing the entire hive.

As with packaged bees or nucs, sourcing bees from an established colony should be done locally. Reach out to your local beekeeping organization or county extension office.

Most suppliers will begin taking orders in the winter. When ready to order your bees, don't wait until spring, or it might be too late. Suppliers only have so many bees available and a small window for shipping. Bees are usually sent out during April or May. Once June hits, it's too hot to ship them safely.

Catching a Wild Swarm

A swarm is a large group of bees that leave their original colony in search of a new home, which happens for one of two reasons: either the colony has run out of room, causing about half the bees to leave and find somewhere else to live, or they abscond, in which

As a new beekeeper, catching a swarm isn't recommended, but an option for those who like a challenge

case the entire colony abandons the hive. Swarming happens most frequently in the spring or early summer.

Clever worker bees know when it's time to swarm and will begin preparing for it. To allow the queen to fly better, they help her slim down by reducing her feeding. Workers will also run around the queen, which causes distress and prevents her from laying eggs.

When a new queen has been reared and is ready to emerge, the original queen bee will then fly off with half the worker bees. Landing somewhere nearby (100–200 yards from the original hive), the swarm forms a cluster as scout bees search for a new hive location. This temporary nomadic state is ideal for catching a swarm.

As you may recall from my own story, I fell into beekeeping when a swarm appeared on my property. Although an exciting and free way to source bees, catching a swarm as a new beekeeper isn't something I'd recommend right out of the gate. For some of the more adventurous folks out there, however, it's certainly an option!

When catching a wild swarm, be aware that just because you've spotted one doesn't mean you can have it. If the swarm isn't found on your property, removing it could be considered stealing under state law. In most cases, landowners will gladly let you have the swarm, but it doesn't hurt to ask for permission first.

To avoid any misunderstandings, it's common to use social media to advertise swarm removal services. It's a great way to get people in your area to reach out voluntarily.

It may not be obvious, but collecting a swarm is actually low risk when it comes to stinging. Before setting out, these brave colonizers fill up on honey for the journey, meaning their bellies are happy and full. Don't we all tend to be more easygoing after a good meal?

Still, having the right equipment and tools to properly deal with a swarm is essential. If you find the bees clustered on a tree limb, learn from my mistake, and *don't panic*. Be gentle, and they'll follow suit in your efforts to relocate them. Here are some step-by-step instructions on how to catch a while swarm:

MATERIALS/EQUIPMENT

- protective clothing
- queen catcher cage
- breathable wooden or cardboard box
- light-colored bedsheet or tarp
- bee brush
- pruning shears
- spray bottle filled with either water or sugar water
- smoker
- lemongrass oil
- packing tape or mesh swarm bag

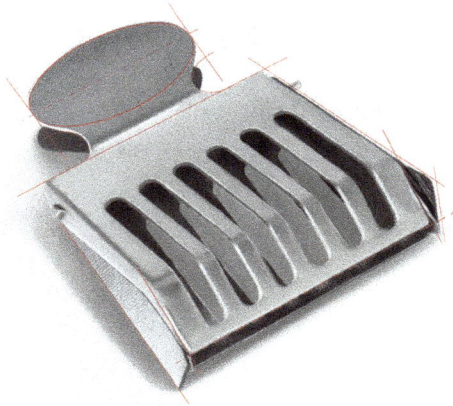

Use a queen catcher cage to isolate the queen and place her in a box for the rest of the swarm to follow

INSTRUCTIONS

Once on site, decide whether it's safe to reach the swarm. If you can access the cluster from ground level, go for it. If it's too high, you'll need a ladder. It probably goes without saying that if it's too darn dangerous to reach the bees in the first place, don't risk it. There will be others.

When you've determined that it's safe to move in on the swarm, get on your protective gear (wearing a hat, veil and gloves is a bare minimum) and take note of the following:

1. If the queen is visible, use a queen catcher cage to attempt isolating her. By placing the cage in your box, the queen's pheromones will help guide the rest of the bees to where you want them.

2. Beneath the swarm, put down a light-colored sheet, placing your box on top.

3. **If the cluster is on a branch**, shake the bees into the box. If hanging from small branches or vegetation, use pruning shears and place the branches in the box along with the bees. The part where you run away screaming is optional.

   If the swarm is on a flat surface or fence post, mist the bees with either water or sugar water so they're less likely to fly away. You can also encourage them to move toward your box by puffing smoke behind them. When ready, lightly brush the bees into the box with a bee brush or piece of cardboard just like you would when using a dustpan.

   If the cluster is on the ground, turn your box on its side and use lemongrass oil to lure the bees in.

   In any of these cases, if the queen hasn't yet been isolated, she'll likely be found near the cluster's center. Don't be surprised if you notice worker bees leaving your box to return to the queen. If this happens, keep at it until the bees stay.

4. Once the swarm has been collected, close the box, leaving a small opening for any stragglers or returning scout bees. Leave the box like this until sundown, as scout bees can spend all day looking for a new hive location.

5. After sunset, seal the box securely with packing tape or place it in a mesh swarm bag. Carefully move the swarm where it can be kept safely overnight.

6. The following day, install the bees in their new home in the early morning to prevent overheating. If there's any remaining vegetation from where you originally sourced the swarm, be sure to remove it. (The "Activating Your Hive" section below outlines how to install packaged bees but is still a good point of reference for installing a freshly caught swarm.)

Once you've managed to get through all these steps, give yourself a pat on the back and celebrate accomplishing something that not many people out there have the guts to do!

With a wild swarm, know that there's also a risk of disease or poor genetics. Sometimes, the queen is injured, dead or difficult to find.

Activating Your Hive

Once all your boxes are checked (hive on-site, fully equipped, bees sourced), it's time to fill your hive with bees! You'll want to get your bees settled into their hive as soon as possible to minimize disruption.

The following will outline how to activate your hive with **packaged bees**. Remember to wear a veil and appropriate protective gear so your bees don't get inside your clothes. In addition, be sure to have the following on hand:

MATERIALS/EQUIPMENT

- hive tool
- smoker
- small nail
- large rubber band
- spray bottle with sugar syrup
- a gallon (or more) of sugar syrup

Make your own 1:1 sugar syrup by mixing equal parts warm water and granulated sugar. This ratio is calculated by weight (not volume), which mimics the sugar content of flower nectar. To give you an idea, 5 lbs of sugar is about the equivalent of 2.5 quarts of water. (Using the metric system, 1 kg of sugar equals about 1 L of water).

Want to see how you can make your own sugar syrup? Watch this:
https://beekeep.blue/1jM

WHEN YOU RECEIVE YOUR BEES

1. Carefully inspect the packaging for any cracks or tears in the screen. Check that your bees are alive and healthy.
    i. It's normal to see a few hundred dead bees sitting at the bottom of the box, so don't be alarmed. Any more than that, however, could mean that there was overheating during shipping, in which case, contact the seller immediately.
    ii. If all looks well, give your bees a generous spray of sugar syrup (don't be overly generous or you'll drown them).
2. Before installation, keep your bee package in a cool, dark place for a few hours to allow the bees to adjust. Keep them away from temperature extremes, loud noises or other disturbances.
3. Spray your bees occasionally with sugar syrup.

An ideal storage temperature is about 60°F (~15.5°C). Anything above 100°F (38°C) will be lethal. When ready for installation, ensure a reasonable temperature above 65°F (18°C).

WHEN YOU'RE READY TO INSTALL YOUR BEES

1. Once your bees have had some time to settle, gather all your equipment and suit up! Make sure you're fully protected.
2. Give your bees another spray of sugar syrup.
3. Holding the bee package from the sides, carry it to the hive, keeping your hands away from the mesh screens to avoid getting stung.
4. Place the package on the ground in a shaded area near the hive.
5. From the hive's bottom box, remove three to four frames from the center of the brood chamber to make space for your bees. Give the bees another spray of sugar syrup.

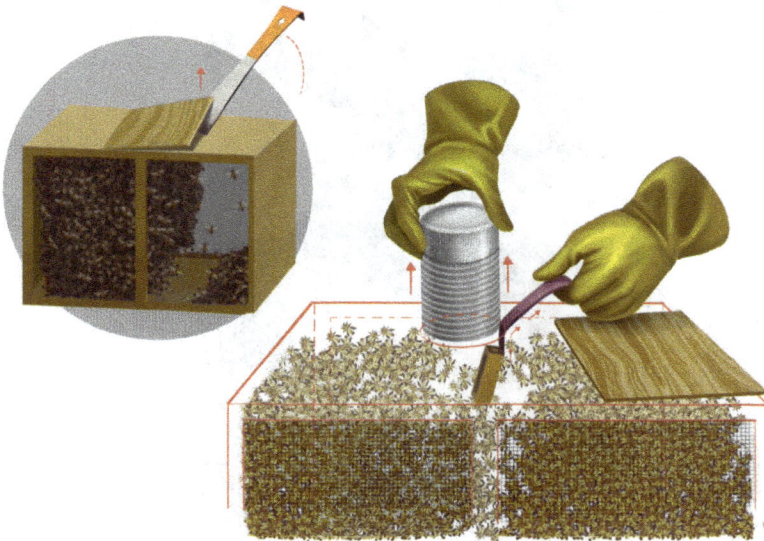

Once you remove the queen cage, check that the queen is alive and healthy

6. Using the hive tool, remove the wooden panel from the bee package. Gently remove the tin feeder and queen cage from the hole at the top of the box. Keep a steady grip on the tab, as this is attached to the queen cage.

7. Replace the wooden panel to prevent any bees from escaping. Carefully shake off any bees that are on the outside of the queen cage.

8. Inspect the queen to make sure she's alive and healthy. Place the queen cage in the shade.

*When loading your bees into their new hive,
give the package a few firm shakes*

LOADING YOUR BEES

1. Hold the wooden panel in place as you knock the package firmly on the ground once, so the bees drop to the bottom.
2. Immediately remove the wooden panel and turn the package upside down over the hive. Shake the bees into the bottom box. It might take a few firm shakes to do this.
3. At this point, there may be bees flying all around you, but don't panic. They're not on the defensive, just disoriented. Eventually, they'll settle down and follow the other bees into the hive.
4. Once you've got as many bees as possible out of the package, prop it up in front of the hive entrance so any lingering bees can crawl in.
5. Gently return all but one of the frames to the hive, taking care not to crush any of the little critters.

INSTALLING THE QUEEN

1. Place a rubber band around the frame where the queen will be installed.
2. Using a small nail, remove the plug from the end of the cage with white candy. Some cages have two plugs, so removing the wrong one could release the queen instantly, which you'll want to avoid.
3. Using the rubber band, secure the queen cage against the foundation a few inches from the top of the frame so the bees can tend to the queen. The cage can also be installed vertically, with the candy end facing up.
4. Return the frame to the center of the hive and gently push the neighboring frame toward the queen cage.
5. **Within a couple of days to a week**, the bees should eat through the candy to release the queen. This slow-release method allows the bees to get used to the queen, increasing the chances of her acceptance into the hive.

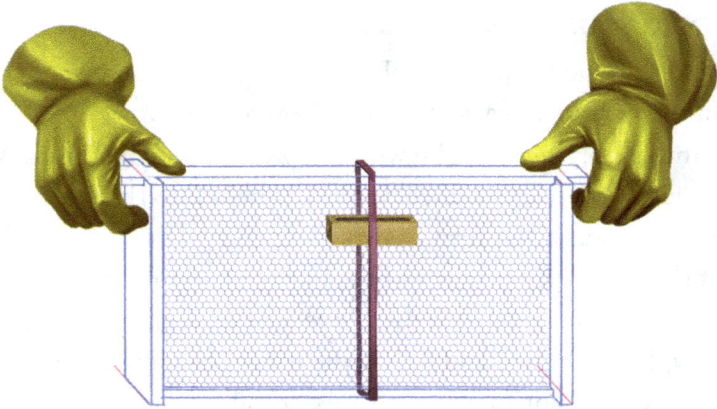

Secure the queen cage against the foundation a few inches from the top of the frame

AFTER INSTALLATION

1. Feed the colony more 1:1 sugar syrup using one of the in-hive feeding methods outlined in Chapter 5. (I recommend opting for a top hive feeder as it's easy to keep full.). Feeding encourages comb building, which is needed to draw out foundation. While adjusting to the hive, your bees will also need a good supply of food before they're able to store honey.

Packaged bees require feeding for around six weeks, so they have time to draw out comb, lay eggs and raise new bees for foraging. When working with a nuc, an established hive, or a wild swarm, feeding over such a long period isn't needed.

2. Once feeding is complete, remember to replace the hive's inner cover and lid. Give your bees time to settle into their new home.
3. **Five days later**, check on the colony to make sure the queen is free from the cage and still alive. Assuming all is well, you can leave your bees to continue doing their thing.

4. Five days after that, check again to see whether the queen has started laying eggs (they look like small grains of rice standing in the center of each cell). This is a good time to add more sugar syrup.

Want to see how to install
packaged bees? Watch this:
https://beekeep.blue/zs0

Common Problems

While the above steps should get your colony off to a flying start, here are some common problems to keep on your radar:

- **The workers reject the queen**. In this case, she'll die in the cage or go missing. Contact your supplier immediately, and they may send a free replacement.

  Another option is to join the hive with an existing colony if you have one. When doing the latter, cover the established hive with a single sheet of newspaper. Poke holes in the newspaper with your hive tool. Place the hive body of the queenless colony on top of the newspaper. The bees will chew through and merge with the other colony.

- **The queen only produces drones**. This situation occurs when she can only lay unfertilized eggs due to improper mating, depleted sperm stores or being subject to the cold before installation. The only solution to this problem is to replace the queen using one of the methods outlined in Chapter 6.

- **A new colony doesn't make enough wax comb**. The solution is simple: feed and feed some more. Bees make comb according to need and require plenty of 1:1 sugar syrup to secrete wax. Even when well-fed, they may need a few weeks to supply enough wax comb for ten frames.

"In 2007, I left the high-tech industry in search of some sanity in my life. A friend called and told me he was taking up beekeeping, which made me think of a poem by William Butler Yeats called 'The Lake Isle of Innisfree.' In it, the lines: 'Nine bean rows will I have there, a hive for the honeybee, and live alone in the bee-loud glade' led me to associate bees with tranquility, resonating with my need for calm in my life. That's how I started beekeeping."

—Stu Farnham

CHAPTER 5
Bee "Having" vs. Beekeeping

The world claims there are three categories of beekeeper that people fall into:

1. **Commercial beekeepers** whose primary source of income is beekeeping. They usually manage a couple hundred hives.
2. **Sideliners** who keep bees and sell bee products as a supplementary source of income. Typically, this describes beekeepers with a few dozen to hundreds of hives.
3. **Hobbyists** or so-called **bee "havers"** who are small-scale beekeepers and tend to re-invest in their bees with any income made. Usually, this refers to beekeepers with 10 or fewer hives.

In beekeeping, the terms "hobbyist" and "hobby" are personally problematic. If you think you can throw a colony in the corner of your yard and expect it to survive without management, it's just not going to happen—at best, your chances are slim. While I'm not suggesting all hobbyists approach beekeeping this way, for me "hobbyist" inadequately describes the commitment required to tend any number of hives, big or small.

When I started looking after bees, it took me some time to understand the importance of treating them as livestock rather than a pastime I could dip in and out of when the mood struck. You wouldn't get cows, horses or chickens and just expect them to fend for themselves. Without actively managing your colonies, you're setting your hives up for failure. What separates real beekeepers from bee "havers" is the time, care and effort put into the practice.

In the spirit of true beekeeping, this chapter will walk you through how to set up your hive, do an inspection and supplement your bees' feeding.

Establishing Your Hive

Your hive will start with a single brood chamber/box at the very bottom. Once multiple frames have been drawn with comb, it's time to add another box. In your first year of beekeeping, your efforts should focus on building your colony up to a double-deep brood chamber (i.e., two brood chambers) with an upper box filled with honey for the winter.

Before going any further, let me explain how drawing comb works. In their natural habitat, bees construct their comb according to their needs and preferences. Comb in the wild is built vertically, from the top down, unlike the horizontal construction in managed beehives.

When using packaged bees to start your colony, they first need to draw comb, which is what naturally occurs when in contact with

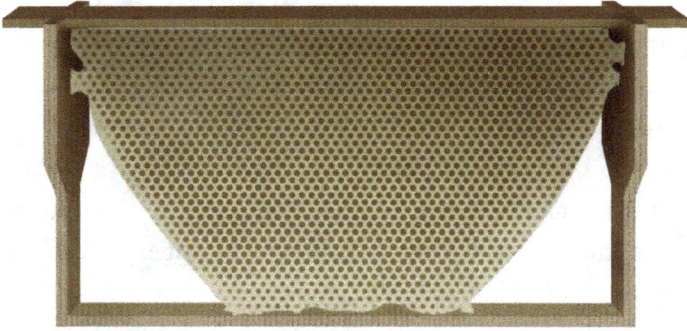

Comb hanging down from a foundationless frame

a foundation. If your frames are foundationless, the comb will hang down from the top of the frame.

In a Langstroth hive, appropriate bee space is maintained by confining comb construction to the foundation enclosed within the frames. With the frames positioned parallel to one another, bees draw comb "out and up," meaning they work their way out from the center of the box until all frames are occupied.

Cross comb can happen when there are large empty spaces in a hive, allowing bees to build in any direction and connecting comb sections between frames. These circumstances make it messy and difficult to remove frames without damaging the comb or harming your bees, which in turn can make them more aggressive.

When working with a nuc, your colony will have already drawn comb. In this case, it's essential to feed your bees enough sugar syrup so they can produce wax and keep up with comb production. If they aren't fed enough, your bees will soon feel overcrowded and likely swarm (especially if your nuc is strong)—not something you want to deal with right off the bat!

Managing Brood Chambers & Honey Supers

With a ten-frame box, opinions differ as to when to add a second brood chamber. For me, the sweet spot is when six or seven frames are drawn with comb: bees continue to draw out comb in the lower brood chamber but have room to expand upwards into the second box.

As the brood nest grows, your bees will eventually focus on building up honey stores. It's then time to add a honey super when seven out of ten frames of the second brood chamber are full. Providing enough space for both brood and honey allows bees to continue their work uninterrupted. If your colony doesn't have enough space, swarming is likely. Leave it too long, and it's almost guaranteed.

People often ask why you wouldn't install all hive boxes from the get-go. It's a fair question and would seem logical, but the problem is that just as cramped conditions are an issue for bees, so is too much space, making it difficult for bees to keep warm in winter and regulate humidity in summer. Bees need to stay warm, and too much space leads to heat loss. Even in the summer, temperatures can drop a lot overnight. Keep your hives the right size for your bees, and only add boxes when needed. Doing this also discourages pests and other insects from moving in.

Inspecting Your Hive

As a diligent beekeeper, you'll need to inspect your hives regularly. Roughly **every ten to twenty days in spring and summer** is a good routine. Inspecting your hives more than once a week will disrupt your bees and set them back a day.

An inspection involves opening your hive, smoking and removing the stacked boxes until you've reached the bottom layer. After carefully inspecting the frames in each box, jot down your observations before putting the hive back together.

Some states have a yearly apiary inspection program, which is meant to uphold hive health and safety standards. While I have my own qualms about the experience and qualifications of county inspectors, be sure to check your local county laws and regulations for more information.

Below, you'll find a detailed guide on how to conduct an inspection. Feel free to refer to each section at your convenience. To get started, be sure to have the following:

MATERIALS/EQUIPMENT
- protective clothing
- smoker
- smoker fuel
- matches or lighter
- handful of green grass
- hive tool
- bee brush
- notepad

Lighting Your Smoker

Your smoker is the most important tool for lowering the risk of getting stung. Smoke disrupts the usual defense mechanism of bees, as their instinct is to move away rather than toward the source.

If you think lighting a smoker is easy, think again. Many beekeepers find that their smoker has gone out by the time they've finished one hive and must go through the trouble of lighting it again to conduct further inspections. Taking the time to master this skill will save you time and frustration in the future.

A smoker works on the same principles as any fire. Think about lighting a campfire: fast-burning fuel like paper, fire starters or tinder is first set alight. As the starter fuel burns, larger bits of wood (or kindling) are added. Once there's a nice fire going, this is when you would pile on some logs.

Conversely, if the logs are placed at the bottom with lighter materials on top, the fire would start quickly but burn out just as fast. Remember, heat rises, so a fire must be built from the bottom up. With a smoker, the problem is that its design often makes it hard to fill the canister and light it from the bottom.

THE TRICK TO LIGHTING A SMOKER

1. Fill the smoker with quick-starting fuel such as crumpled paper, wood shavings or pine needles. Make sure the fuel isn't too densely packed so there's space for oxygen to circulate.
2. Light the fuel. When it begins to burn on its own, use your hive tool to push it to the bottom of the smoker. Pump the bellows a few times so air is forced up through the fuel.
3. Once the first lot of fuel is burning nicely, add more of the same. As it burns, use the hive tool once more to push it down, and again, pump the bellows.
4. Repeat steps 2–3 several times. As the new fuel starts burning, push it to the bottom and ventilate.

crumpled   **wood**   **pine**
paper   **shavings**   **needles**

5. When you've got a good fire going with flames coming up the fuel chamber, larger pieces of fuel and more oxygen can then be added.

6. Before closing the lid, add some green grass to the fuel chamber, which helps keep the smoke cool as it leaves the smoker—hot smoke can scorch the bees' wings!

7. Close the lid as the larger pieces of fuel begin to burn. From this point on, the fire should happily smolder on its own. All that's needed is an occasional pump on the bellows.

8. Keep an eye on fuel levels so you can top up if needed. Whenever fuel is added, give the smoker a few more pumps of air.

For all you visual folks, I break down how to light a smoker in this video:
https://beekeep.blue/4dx

A smoker will stay hot for a good while when it has a strong flame. This can be easy to forget after you've finished your hive inspection, so remember to keep your smoker somewhere safe as it cools down.

Plugging the tip of your smoker after finishing up will extinguish the hot coals faster and also save unburned fuel for the next inspection.

Opening The Hive

Inspecting your hive will no doubt require you to open it up. At this stage, all your protective gear and equipment should be ready to go. Always approach your hive from the side or rear—walking in front of it will obstruct your bees' flight path.

Using more smoke than necessary actually makes your bees work harder. Contrary to belief, it won't make them any calmer but encourages them to consume stored food as they prepare to evacuate the hive as if it were on fire.

AS YOU APPROACH THE HIVE

1. Observe your bees as they leave the hive. They will usually fly straight out, but you may notice them flying to the right or left, in which case, approach the hive from the opposite side.
2. Keep your smoker about two feet away from the hive entrance. Blow about four puffs of smoke toward the hive—no need to over-smoke your bees, as you're just letting them know you're there.

Smoke disrupts the usual defense mechanism of bees

REMOVING THE OUTER COVER

1. As you stand to the side of the hive, lift the long edge of the outer cover an inch, blowing a few more puffs of smoke into the hive. Replace the cover.
2. The smoke will alert any guard bees hanging around the top of the hive. Wait 30 seconds for the smoke to work its way in.
3. Put your smoker down and use both hands to slowly remove the outer cover, lifting it straight up. Flip the outer cover over and place it on the ground.

REMOVING THE INNER COVER

1. Give the hive another few puffs of smoke through the hole of the inner cover.
2. Starting with one side, gently press down on the inner cover with one hand while using the hive tool to pry it loose with the other. You may hear a loud snapping sound when they come apart, which can startle your bees. Move slowly.
3. Walk around the hive to do the same on the opposite side.
4. With both sides now loosened, use your hive tool to pry the inner cover up and blow smoke into the gap created. Wait 30 seconds before removing it entirely.
5. Take care not to crush any clinging bees as you prop the inner cover up against the hive.

It's a misconception that doing inspections at night is less disruptive to your bees. The exception is some of the more aggressive breeds (e.g., Africanized honeybees), in which case inspections may be more manageable after dark. In these cases, using a red light is helpful when harvesting honey and inspecting frames.

WORKING YOUR WAY THROUGH THE HIVE

1. With the inner cover removed, use your hive tool to scrape off any wax or propolis found at the top of frames or around the hive walls.

2. It's now time to pry up the honey super. Remove it with your hive tool and gently place it aside. If your hive has a second honey super, smoke and remove it too. If applicable, you'll also remove the queen excluder.

3. Next, carefully puff smoke into the top brood chamber (also called "the second deep")—in most hives, it's one of two boxes containing the brood colony. Wait a few minutes for the smoke to take effect, then remove it for later inspection, placing it on top of the super or outer cover.

4. Finally, smoke the bottom brood chamber.

Inspect the frames in the order they are found

5. After a few minutes, the inspection can begin. As you go, keep the frames in the order they're found. If needed, use a bee brush or extra smoke to move any bees out of the way. Start by removing the first frame (which will most likely still need drawing out). Gently place it on top of the other hive boxes, always being mindful of any clinging bees.

6. To avoid pinching or rolling the queen, use your hive tool to gently move frames aside before lifting and inspecting each one.

Sometimes, bees can get pressed between frames. When a queen gets pinched between this space or between a bunch of bees, she can get rolled when the frame she's in or a neighboring frame is lifted. These situations can hurt or even kill the queen.

WHAT TO LOOK FOR

1. First, try spotting the queen. Identifying her is easier when she's marked, but if she's not, look for a bee with a long, slender abdomen surrounded by a circle of workers.

2. If you can't find the queen, look for eggs, which at least indicate she's been around the past few days.

3. Next, check for parasites or pests. **Deformed Wing Virus (DWV)** is easy to detect—the name is pretty descriptive too— it's also a sign that your bees are likely affected by Varroa mites.

4. During your first hive inspections, it's important to count how many frames are drawn out or filled with comb. When seven of the ten frames are drawn out in the bottom brood chamber, you'll know you need to introduce the second brood box.

Eventually, when seven of those ten frames are full, it's time to add a honey super. And, during a subsequent inspection, when the first honey super is close to full, you'll put on another.

*If the queen bee isn't marked, look for a longer and
more slender bee surrounded by a circle of workers*

Brood is a mixture of capped/uncapped larvae and eggs

5. After checking for parasites and pests, it's time to check for brood, which consists of a mixture of capped and uncapped larvae and eggs.

6. When inspecting the eggs, they'll look like thin grains of rice standing on end in the middle of the cell. If you notice more than one egg per cell, your hive has laying worker bees (consult Chapter 6 on how to fix this). You may also want to contact a more experienced beekeeper for support.

For the best chance of spotting eggs, start at the bottom center of the frame. With the sun behind you, angle the frame toward the sky at roughly 30°. Ensure your veil doesn't cast a shadow and obstruct your view by holding the frame slightly to the side. Using reading glasses or a magnifying glass can also be helpful.

After an inspection, it's time to put the hive back together

7. Lastly, check for food stores. I look for a band of honey or nectar across the top of brood frames. If there's little to no honey found, it's a sign that your bees need feeding.

8. After each inspection, gently return the frame to its place. When you're done with the last frame, push all the frames together with your hive tool to make space at the front for the first frame to be returned. Ensure the frames are centered in the box.

9. Once you've finished inspecting the bottom brood box, repeat the inspection process with the top brood chamber.

See what to look out for during a hive inspection:
https://beekeep.blue/Rfn

PUTTING IT ALL BACK TOGETHER

1. You can now return the top brood chamber to its place above the bottom box.

2. If your hive has a queen excluder, that will follow before putting the honey super back in place. Align the front edge of the super with the hive's back edge, moving carefully as you push it forward. As needed, use your smoker or bee brush to remove any bees.

3. Next, slowly slide the inner cover back in place across the honey super. Smoke or brush bees away as needed. Finally, gently return the outer cover.

4. As soon as you're done, write any observations down in a notepad or journal to keep track of your hive's health. For any basic troubleshooting needed, consult Chapter 8.

5. Remove your protective clothing and put your smoker somewhere safe to burn out.

Feeding Your Bees

As livestock, bees need feeding. Supplementing their diet throughout the year can be vital to their survival, especially during winter and early spring. In this section, I'll share what to consider when feeding your bees with the changing seasons.

Spring Feeding

Spring is a critical time for bee colonies as they start to increase brood production and build up their population. Early in the season, check for food stores without opening your hive, which preserves warmth. By placing one hand under the bottom board and gently lifting, you can assess your hive's weight—it's a skil that might not come so easily at first but is quickly learned with time and experience.

If the hive feels heavy, there's still plenty of food. If the hive feels light, however, it's an indication that your bees have worked through most of their food stores, and you'll want to give them a boost while waiting for flowers to bloom.

Once feeding starts, continue until nectar becomes available, which will all depend on local weather conditions. During nectar

When nectar becomes available, foraging activity naturally increases

flow, there is an increase in foraging activity and bees are relatively easy to manage, as food sources are plentiful. For this reason, there's little to no hive robbing. You should also notice an increase in stored nectar.

Aside from timing it with the flow of nectar, you'll also want to stop feeding before any supers are added to your hive. Otherwise, bees will move sugar syrup into the super, contaminating the honey.

POLLEN PATTIES IN THE SPRING

Pollen substitutes, such as pollen patties can be purchased from beekeeping supply stores or prepared at home using commercially available pollen substitute powders. Pollen patties are high in protein and other essential nutrients, which promote brood rearing. Since having more bees means a stronger workforce to forage for nectar, this is a supplement used by many beekeepers.

Rule of thumb for beginners: don't offer pollen substitutes until pollen becomes available in nature. Boosting brood production too early in the spring season is for the more experienced beekeeper, as bees can quickly run out of room and decide to swarm.

To feed your bees pollen patties, place them directly on top of the frames inside the hive. Ensure that the patties are easily accessible and are not obstructing hive entrances or ventilation. Proper ventilation is essential to prevent excess moisture buildup in the hive.

**Watch how to make
pollen patties here:**
https://beekeep.blue/Fhp

SUGAR SYRUP IN THE SPRING

As temperatures slowly rise above 40°F (~4.5°C) overnight, it's safe to feed your bees 1:1 sugar syrup. Refer to the "In-Hive Feeding" section for more information. Remember, this ratio refers to weight and not volume.

Summer Feeding

In most regions, summer is a period of abundant natural forage for bees, including flowers and plants from which they gather nectar and pollen. For this reason, providing supplemental feed to your bees during the summer is generally not necessary.

Exceptions to this may include an unusually early summer drought or a lack of flowering plants in the area. During periods of dearth or when environmental conditions are unfavorable for foraging, supplementing your colony's food stores can help prevent starvation and maintain hive strength.

Additionally, newly established colonies or weak colonies might need feeding to help them build up. Monitoring hive conditions and weather patterns in the summer is crucial to determine if/when feeding is necessary.

If using **sugar syrup** as supplemental feed, regularly check hive feeders to ensure they aren't empty, especially during periods of high bee activity. Remember to monitor hive weight to gauge honey stores and determine if supplementary feeding is needed. **Pollen patties** are also a good option to give your colony that additional boost.

Fall Feeding

Even if they've brought in some fall honey, feeding during this time of year is crucial to helping bees survive the winter. If a colony doesn't have enough food stored, it will starve. Bees need plenty of food to maintain the temperature of their winter cluster. They create warmth by constantly vibrating their flight muscles, which requires a lot of energy.

Fall feeding must be done before temperatures drop below 50°F (10°C), which is when bees stop flying. A good time is after you've had a chance to harvest honey. Worker bees need enough time to cure and cap the sugar syrup so it can be stored for winter. Where I live in the Midwest, colonies should be fully fed by mid-November, although to be safe, you might want to do this earlier.

As you head into winter, a hive with a single brood chamber should weigh between 70–90 lbs. A hive with a double brood chamber should weigh around 100–120 lbs. As described above, weighing your hives isn't necessary once you have a reference point—simply place a hand under the bottom board to see if you can lift the hive from one side. If it's ready for winter, it will feel heavy. The more you practice this habit, the easier it will become to assess.

SUGAR SYRUP IN THE FALL

In the fall, bees need a thicker syrup with a 2:1 sugar-to-water ratio. The higher sugar content means it can be ripened quickly for storage. If it has too high a moisture content, this can lead to dysentery, a common cause of overwintering colony death.

A colony needs at least 4 gallons or 15 L of sugar syrup in the fall. You can always give more if your bees continue to feed. A hive with a single brood chamber should weigh between 70–90 lbs as you head into winter.

POLLEN PATTIES IN THE FALL

Offering your bees pollen patties in the fall allows your colony to grow its population as it moves into winter. While having more bees to cluster and keep warm during the cold season is an advantage, it can be a double-edged sword, as this also means having more mouths to feed. Uneaten protein patties can also attract pests such as wax moths and small hive beetles, so be sure to monitor your hives carefully.

DRY SUGAR/CANDY BOARDS (SUB 40°F)

When overnight temperatures are lower than 40°F (about 4°C), dry sugar or a candy board can be used to supplement feeding.

For dry sugar feeding, use the mountain camp method: place a sheet of wax paper or newspaper on top of the frames in the brood box. After poking a few small holes around the outer edge, pour dry sugar in the middle of the sheet.

Bees will only eat dry sugar if it's cold outside! You see, bees maintain the temperature of the hive's core, where the cluster forms, at around 95°F (35°C) even during winter. The temperature gradient between the warm hive interior and the cold exterior causes moisture to form inside the hive. When bees respire, moisture is released into the air, which condenses on the sugar, making it slightly damp and easier to consume.

Check out this video on
the mountain camp method:
https://beekeep.blue/bkV

A candy board is made by mixing granulated sugar with water to form a thick paste. The sugar paste is then set and hardened in a frame or mold. Place the candy board directly on top of the frames inside the hive.

As temperatures drop below 30°F (-1°C) later in the season, however, the sugar tends to crystallize, making it harder for bees to consume.

**Watch how to make a
candy board in this video:**
https://beekeep.blue/CkG

Winter Feeding

In winter, honeybee colonies rely on stored honey and pollen to sustain themselves when foraging opportunities are limited. Regularly monitoring hive weight and food stores throughout the winter will determine if emergency feeding is necessary. Do your best to help your bees survive the winter by supplementing their feed before it gets too cold.

DRY SUGAR/CANDY BOARDS (SUB 40°F)

Dry sugar or a candy board can be used as a long-lasting food source throughout the winter. Place it directly inside the brood chamber so your bees have easy access. The advantage of using dry sugar is that it absorbs moisture in the hive, keeping bees nice and dry. When it gets too cold, however, crystallized sugar is not easy to consume.

WINTER PATTIES

To help them survive winter, bees can also benefit from winter patties, which are different from pollen patties, as they contain little protein and are mostly packed with carbohydrates and other nutrients. Winter patties can support your colony's health without causing a rapid population increase that would dwindle food stores when your bees need them most.

It's important to compare feeding schedules to other beekeepers in your specific climate!

	winter patties	pollen patties	dry sugar	sugar syrup	fondant / candy board
WINTER no higher than 40°F / 4°C	X		X		X
SPRING no lower than 40°F / 4°C		X	X	X	
SUMMER above 55°F / 12.7°C		X		X	
FALL no lower than 55°F / 12.7°C		X	X	X	X
NOTES	<ul><li>Temperatures shown are overnight</li><li>Pollen subtitutes of any kind will encourage bees to raise brood. Brood rearing shouldn't be encouraged when temperatures are below 30°F / –1°C</li><li>(x) Dry sugar can't be consumed by bees unless it contains or absorbs moisture. It is usually fed when overnight temperatures are below 40°F / 4°C (in winter/early spring/late fall)</li><li>(x) Sugar syrup can be fed when outside temperatures are above 40°F / 4°C overnight</li></ul>				

Seasonal feeding guide for Central Ohio

In-Hive Feeding

When it comes to supplementing your bees' diet, it's best to do "home delivery," meaning there's no need to leave the hive. Some beekeepers place food outside the hive, which is referred to as open feeding, but this can lead to all kinds of problems like contamination or robbing. It's not a method I would recommend for new beekeepers.

Dry sugar, candy boards, pollen or winter patties are all placed directly inside the hive on top of frames. When using **sugar syrup** as a supplemental food source, there are a variety of ways you can go about in-hive feeding.

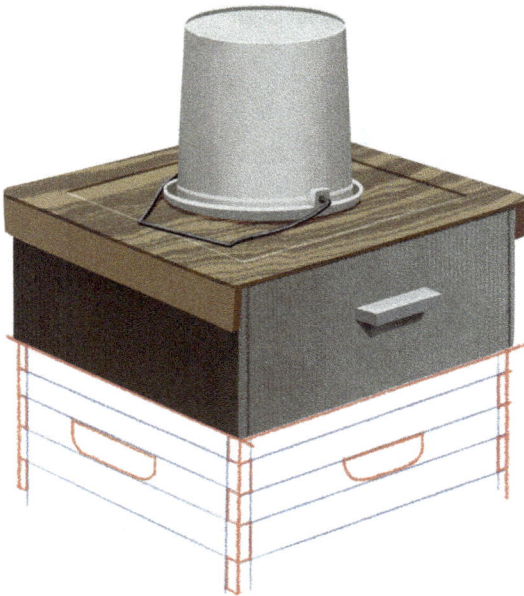

You can purchase a bucket with a screened lid or poke holes in a regular lid

Bucket Feeding

MATERIALS/EQUIPMENT

- bucket with a screened lid
- inner cover with a hole at the center
- sugar syrup
- weights, such as rocks or bricks

If making your own screened lid:

- marker and ruler
- finishing nail

INSTRUCTIONS

1. If making your own screened lid, trace a circle about 3 inches in diameter at the center of a regular plastic lid. Staying within the marked circle, poke a finishing nail through the lid to make holes.
2. Fill the bucket with sugar syrup and cover with the modified lid. Turn the bucket upside down so it rests over the hole of the inner cover.
3. Secure the bucket in place with weights like bricks or rocks.

Hive-Top Feeding

MATERIALS/EQUIPMENT

- hive-top feeder
- sugar syrup
- straw or dry grass
- heavy-duty tape (*Gorilla Tape works wonders!*)

INSTRUCTIONS

1. Place the hive-top feeder directly over the brood box.
2. Carefully pour and fill the hive-top feeder with sugar syrup. Any spillage on the outside of the hive could trigger hive robbing.
3. Replace the inner cover.

4. Using heavy-duty tape, patch any gaps or holes around the feeder where pests like wasps could get in.

Bees can access the syrup by climbing up the feeder chimney

Baggie Feeding
MATERIALS/EQUIPMENT
- resealable plastic bag
- sugar syrup
- knife or nail
- wooden spacer shim

INSTRUCTIONS

1. Fill ⅔ of a large resealable plastic bag with sugar syrup and place it on the top bars of the hive.
2. Using a knife or nail, poke a few holes in the bag so bees can access the syrup.
3. Place the wooden shim around the bag before putting the inner cover back into place—this prevents the bag from getting squashed and spilling syrup all over the bees.
4. Refill as needed.

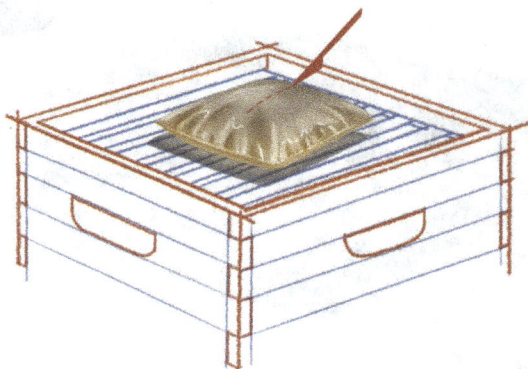

Baggie feeding is a cost-effective method

Frame Feeding

Since the brood chamber needs to be opened to install the feeder when using this method, only frame feed when temperatures are above 50°F (10°C). Make sure the frame feeder goes empty or that it's removed after your bees are done feeding. If not, the frame feeder will add unnecessary moisture to the hive, which can be lethal for a colony. Remember to replace the feeder with the original frame.

MATERIALS/EQUIPMENT

- frame feeder

INSTRUCTIONS

To install a frame feeder, the instructions are simple:

1. Remove one of the empty frames from the brood chamber closest to the brood nest and replace it with a syrup-filled frame feeder.

A frame feeder sits inside the brood box

PART III

Becoming a Seasoned Beekeeper

"I started beekeeping when my kids and I learned about it in our homeschool curriculum. In recent years, I've been thinking about the pollinator shortage and felt it would be a good way to help. The kids and I got our first hive last year and quickly became obsessed. We now have nine hives, and I love every minute I'm with our bees. They calm me, and in a way, are a sort of therapy."

—Karen Teague, Fuzzy Nuggets Apiary

CHAPTER 6

Growing & Maintaining Your Colony

As responsible beekeepers, it's our duty to support our bees in their regular hive activity by minimizing stressful situations and providing the proper care and attention. With the changing seasons in the Midwest, it's important to address the relevant beekeeping tasks that should be carried out year-round.

After building your beekeeping momentum and establishing your first hives, you'll begin to notice a regular pattern in your bees' behavior. With this, it's important to recognize different queen cells and how to manage each situation accordingly.

A crucial part of this process is learning to introduce a new queen to a hive. This same procedure applies when a current queen bee is at risk or your colony is suddenly left queenless.

Everything covered in the following sections is essential to growing and maintaining a thriving colony as you continue to develop your beekeeping skills.

Seasonal Changes & Tasks

Changes in weather have a direct effect on hive activity: the temperature, wind, rainfall and sunlight all determine when and if honeybees can fly. Flying means foraging, which is when bees search for nectar, pollen, water and propolis to sustain the colony. As a result, more flying hours means more honey. Of course, the weather also makes a difference to the local flora and crops. Wet, dry, cool or warm conditions all influence the season's nectar flow. A short spell of foul weather can significantly impact that year's honey supply.

Throughout the seasons, here are some important points to consider:

Late Winter to Early Spring

During this period, the queen lays eggs and brood rearing begins. Nurse bees tend to the brood and feed them a mixture of bee bread, royal jelly, pollen, honey and nectar, all depending on the brood's stage of development. When temperatures are above 50°F (10°C) and the weather is fine, worker bees begin foraging for nectar and pollen.

This seasonal transition can be a challenging time of year for bees. With the high energy demands that come with brood rearing and unstable weather conditions ranging from sunshine to freezing snaps, the colony may quickly exhaust its resources.

From late winter to early spring, your job is to maintain the hive's temperature and ensure your bees have plenty of resources to feed their young. Inspect your colonies regularly without removing boxes or frames. Since bees move up as winter progresses, knowing where they are in the hive will indicate whether there's still food available above them. Pop the top of your hive open and look down through the frames to confirm the location of your bees. As shared in chapter 5, you can also assess your hive's weight by gently lifting from the bottom board.

This time of year is also when brood diseases are most likely to hit, which is why focusing on your hive's health is crucial (see Chapter 8). Many brood diseases are easily treated with extra feeding and improving access to floral resources. If your colony is too weak to survive, you may need to make the difficult decision to cull it, rather than have the disease spread to healthy colonies.

On the bright side, once flowers and trees are in bloom, there is an abundance of nectar and pollen available. Spring is also the season for swarming, which makes it a good time to establish a new colony.

LATE WINTER/EARLY SPRING TASKS

- Keep the entrance of your hive cleared from dead bees and snow/ice.
- Remember to check in on food stores and supplement when needed. Feed your bees until blooming flowers are available (refer to Chapter 5 for more details).
- Have one or two empty hives ready in case of a swarm. Otherwise, you might lose half of your bees when they set off to find a new home.
- Harvest any honey left unused after winter (more on that in Chapter 7). If you fed your bees in the fall, don't harvest at this time, as it will be sugar syrup and not honey.
- Inspect the hive for a healthy brood pattern, which should look uniform as it moves outward from the center of the frame. A spotty brood pattern with many empty cells in between isn't a good sign. If you suspect the queen has died, she'll need to be replaced.

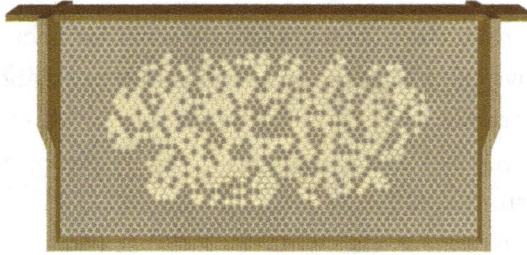

spotty brood pattern

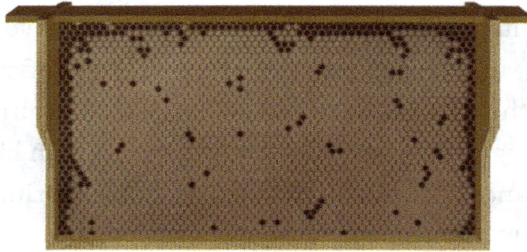

solid brood pattern

- Checkerboard your hives to delay or prevent swarming (see below for more details).
- If you have multiple hives, spread out your bee population so each hive has roughly the same numbers.

Checkerboarding

It's an excellent time to checkerboard your hives in early spring. This technique breaks up the solid ring of honey that surrounds the top of the brood nest, which is what tells the bees that it's time to swarm. Checkerboarding stimulates honey production and can easily be applied when checking for food stores.

MATERIALS/EQUIPMENT

- empty super
- empty frames

INSTRUCTIONS

1. Start by removing alternate frames of honey from your hive's super and replace them with frames of empty comb or foundation.

2. Place the honey-filled frames in the empty super, keeping the same position they were in before. Both supers should now have alternating full and empty frames. This arrangement exposes more storage areas, which delays swarming until they're all filled. There's a chance that swarming is put off entirely.

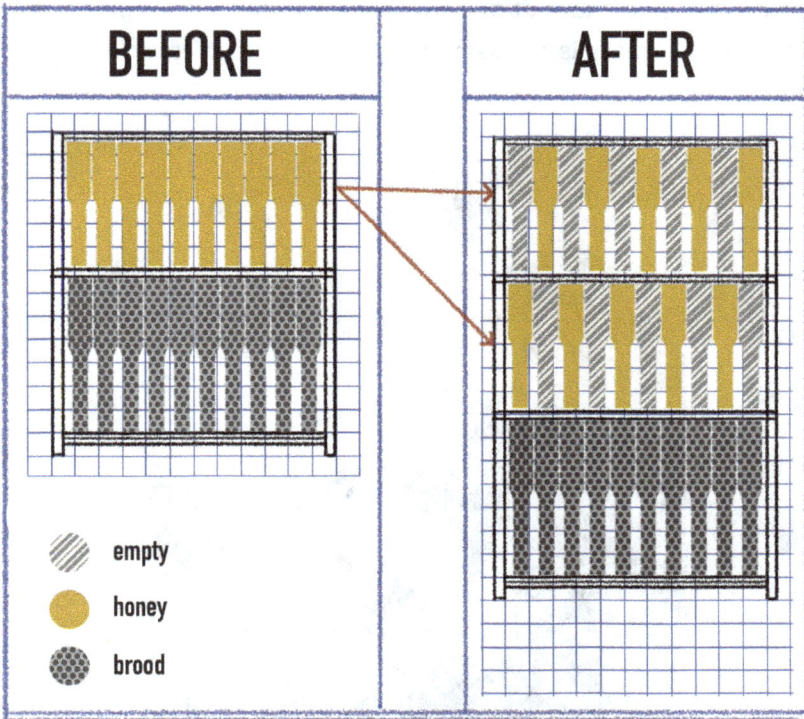

Checkerboarding: alternate full and empty frames to stimulate honey production and delay swarming

Late Spring to Early Summer

By now, your colonies will have been actively rearing new bees for several weeks and floral resources will be readily available, enabling the colony population to expand rapidly. This growth triggers the rearing of new queens and drones.

New queens are reared when the level of queen pheromone drops. This chemical helps bees identify their queen and signals that she is present. In a larger, denser colony, there's less of the queen pheromone throughout the brood nest. Nurse bees usually rear queens on the edge of frames where pheromone levels are lower. The appearance of many queen cups will be your first visible cue that swarming is a possibility, so it's wise to keep a close watch on the colony.

*Queen cups are built as a preventive measure and
turned into queen cells when necessary*

When raising a new queen, worker bees will expand the queen cup into a larger queen cell, where the queen larva is fed a special diet. In the event that a new queen isn't required, the queen cups may remain empty or be used for storing nectar.

When a colony swarms, around half the bees will leave along with the old queen. The process for a new queen to begin laying eggs can take up to four weeks: 16 days for the queen to emerge, seven days to mate, and another seven days to start laying.

While swarming can be put off with regular hive management, it's also useful to look out for queen cells. Although some beekeepers destroy developing queen cells to prevent swarming, this is time-consuming and doesn't always solve the problem. A vibrant overwintered colony will still try to swarm, and if resources are abundant, colonies started from nucs or packages may also swarm. A better approach is to split the colony as outlined in a later section.

LATE SPRING/EARLY SUMMER TASKS
- Check for queen cells (details provided in the following section).
- Since it takes around 16 days for a new queen to develop, inspect hives at least once every two weeks so you can add extra brood frames or honey supers to delay/avoid swarming. Remember the goal here is to make sure the queen always has open cells to lay in in the brood box
- Split your colonies to prevent swarming.

Moving Into Late Summer

During summer, colonies build honey stores for fall and winter. If you split your colony or it swarmed, new queens will have mated and started to lay eggs. This period is when many beekeepers harvest honey.

It's also important to start checking for Varroa mites—one of the biggest threats to honeybees. When foraging, bees mingle with other bees, which is when mites can be transmitted. Mite numbers also increase quickly with high levels of brood rearing, so you'll find that mites are most likely to affect your biggest, strongest colonies.

As summer progresses, nectar flows decrease, and food becomes scarce. Bees tend to become more defensive, and strong colonies may start robbing smaller or weaker ones. If there's any time to cut back on colony inspections, this is it. You might also like to install a robbing screen across the entrance of your hive to prevent non-resident bees from getting in.

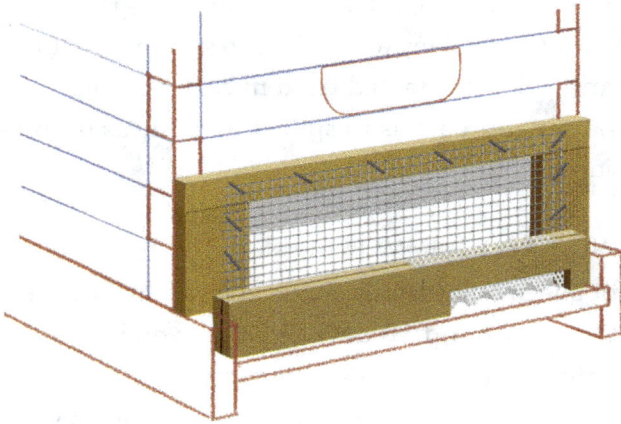

The late summer is a good time to install a robbing screen

**To see how robbing screens work,
take a look at this video:**
https://beekeep.blue/8xH

During this period, colonies will start rearing winter bees, which are physiologically different from summer bees. They live longer (six months instead of six weeks) and have larger, fatter bodies equipped to endure the cold. Healthy numbers of winter bees are vital to the survival of a colony. If they're unhealthy or carry disease, they are unlikely to survive the winter season, leaving the winter cluster too small to make it through to the following year.

LATE SUMMER TASKS

- Make sure your hives have nearby water sources.
- Install a robbing screen to prevent stronger hives from robbing weaker ones.
- Check that your queen has a healthy laying pattern and plenty of room to lay.
- Keep an eye out for Varroa mite infestations by performing an alcohol wash (see Chapter 8). Use other mite treatment when needed and be sure to follow the provided instructions.
- Harvest your honey.
- Remove the supers and start feeding your bees. If you wait until nectar is no longer available to do this, there's a chance your bees won't eat much because of the colder temperatures.

Fall

At this time of year, there's often another nectar flow, sometimes called the "fall flow." This extra flow allows bee colonies to build up food stores for the winter. Although winter bees are still being reared, brood production slows down.

Soon, your colony will be entirely reliant on its food stores. As resources dwindle, worker bees become more defensive. During this time, drones are often thrown out of the hive because they're no longer helpful to the colony and are merely a drain on resources.

You might even observe worker bees carrying uncooperative drones out of the hive.

In the fall, take note of your colonies' weight. The ones with a large population and plenty of honey have the greatest chance of surviving winter. If you notice any weak or light colonies, give them extra supplemental feed or combine them with another small colony, increasing their chances of survival.

FALL TASKS

- Ensure your colonies have enough honey stored up for the winter.
- Check your combs to ensure there's a good, uniform pattern.
- Inspect for disease and treat or dispose of frames where necessary. Continue watching for Varroa mites and monitor the effectiveness of any treatment. Mite populations should be as small as possible to give your bees the best chance of surviving the winter.
- Combine weak (and disease-free) colonies with stronger ones.
- Reduce the hive entrance, install a mouse guard and ensure your hives are properly ventilated.
- Weigh down the tops of your hives to protect them from strong winds.

Winter

A honeybee's winter starts when the frost has killed the last flowers. All summer bees will have died and brood rearing stops. The cold triggers the forming of a cluster, which is when bees gather in a ball shape across multiple frames. The bees vibrate their flight muscles to generate heat and keep the colony warm. This warmth is vital to the movement of worker bees that will eventually leave the cluster to get food. With that said, after the first

Forming a cluster helps bees stay warm during the winter

frost, avoid breaking boxes apart and pulling frames during inspections to preserve the warmth in your hive.

A cluster usually starts at the bottom of the hive and moves laterally and vertically across the hive to consume honey stores. Since bees are unable to consume cold sugar syrup, local beekeepers in areas with particularly harsh winters may place fondant or a candy board at the top of the uppermost frames to feed their bees.

Winter is also an excellent time to lower the mite count in your hives. The lack of brood makes it the perfect opportunity to apply an oxalic acid dribble, killing any mites praying on adult bees. This way, the colony can restart its lifecycle in the spring with little to no mites.

WINTER TASKS
- Complete all disease treatments, including an oxalic acid dribble or vaporization, so that the colony can head into winter with a small mite load.
- Install a windbreak to shelter hives from the wind.

- Regularly inspect your hives for wind damage and ensure adequate ventilation. Sealing the hives entirely causes condensation that can severely harm your colony.
- On warmer days, briefly open the top of hives and have a quick peek to see whether your bees have enough honey for food. If they don't, try the mountain camp method or offer fondant. Remember—refrain from breaking boxes apart or removing frames to preserve warmth.

Want to see how to prepare your bees for the winter? Watch this:
https://beekeep.blue/pcy

When the late winter and early spring come around once again, you can order new materials for replenishing or expanding your hives as late as the end of February/March.

Queen Cells

There are many types of cells in a beehive, including honey storage cells, drone cells, brood cells and pollen cells. Queen cells, on the other hand, are different from all of these—rather than the usual hexagonal shape, they're larger and shaped more like a peanut.

Identifying queen cells in your hive is essential. While some beekeepers will advise you to exterminate any queen cells you find, it's much wiser to figure out *why* your bees are building them before taking action. Are the bees about to swarm? Is the queen having problems carrying out her duties? Has the hive been left queenless?

Depending on the circumstances, there are three kinds of queen cells: swarm cells, supersedure cells and emergency cells. As these

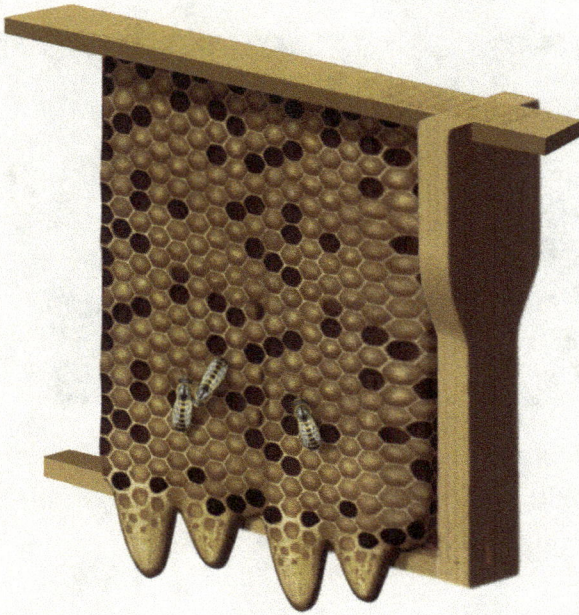

Swarm cells are often found hanging off the bottom of frames

cells are easily damaged, be careful when handling them and keep them vertical when removing frames from the hive.

Swarm Cells

When a colony becomes too big and the queen is preparing to leave with half of the bees to find new hive space, this is when swarm cells appear. It's a sign that you've got a healthy, thriving colony. Swarm cells are used to raise a new queen that will then take over the original hive.

This kind of queen cell usually hangs vertically off the bottom of frames. They can often be mistaken for drone cells (which are rounded and shaped like a bullet). Swarm cells have more of a lumpy peanut or teacup-like shape. You'll usually find ten to twenty

*Unlike swarm cells, supersedure cells are created
closer to the center of the frame*

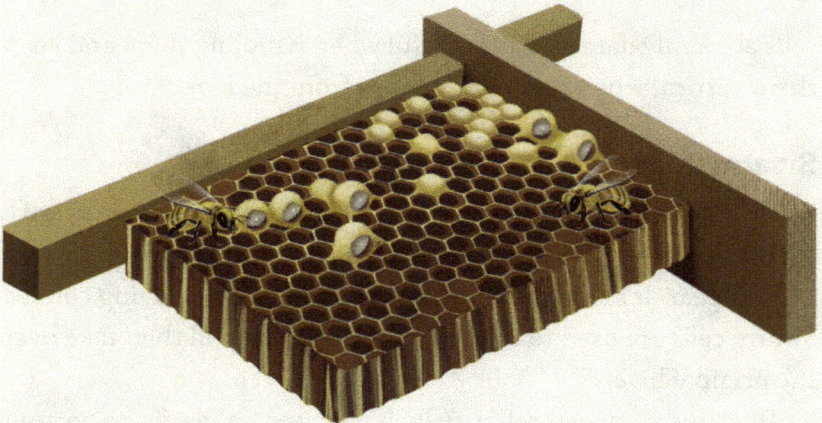

*Emergency cells are built when a hive
is suddenly left queenless*

swarm cells at a time, although sometimes there can be up to a hundred.

Bees tend to swarm when hive populations increase in spring or summer. If you spot swarm cells during this time, your bees are already in swarming mode, and stopping them may not be possible. You can try removing swarm cells before they hatch, but this will only delay what's inevitable.

As mentioned, if you notice swarm cells in your hive, the best thing to do is to split your colony, which mitigates the risk of losing half of your bees in the swarming process. Keep reading for more details on how to carry out this procedure.

Alternatively, you can prevent swarming by adding boxes as needed. The goal here is to make sure the queen always has room to lay and the bees have a place to store food.

Supersedure Cells

Bees can sense when their queen is injured, sick or old and will prepare for her replacement by creating supersedure cells. Unlike swarm cells that are built on the edge of frames, supersedure cells are constructed closer to the center, so they're mostly kept hidden from the existing queen. It's amazing how clever bees are!

Several supersedure cells will usually be created at a time, giving the hive the best chance of raising a strong new queen. Young larvae are fed royal jelly in these cases, and a supersedure cell is then built around each for protection. Usually, the first bee to emerge will end up killing the others so she can claim her title as queen and subsequently take over the hive.

Emergency Cells

If a hive is suddenly left without a queen, bees will go into emergency mode to raise a new one as soon as possible. Nurse bees will convert normal brood cells into emergency cells, and the brood is then fed royal jelly. While it takes around 16 days for a new queen to emerge, if you haven't seen one after that time, consider introducing a queen to the hive yourself by following the instructions provided later in this chapter.

You can identify an emergency cell because it's found partly between a brood cell and draped over the edge of the comb. They're also built anywhere on the comb and may be scattered rather than clustered, making it easy to tell emergency cells apart from swarm or supersedure cells.

Laying Worker Bees

When a queen is dying or dies, workers will focus on raising a new queen, although sometimes, this doesn't work out as it should. Roughly ten days later, the hive is left with no more open brood. Open brood refers to the larval stage before the cell is capped and the larva begins to pupate.

With no open brood left, the open worker brood pheromone fades, causing worker bee ovaries to mature. Gradually, worker bees then begin laying eggs (several per cell), which start to appear around three weeks after losing the queen. Since worker bees cannot mate, these unfertilized eggs later develop into drones.

The biggest issue here is that laying workers also produce enough queen-like pheromones to fool the colony into thinking it has a queen. As a result, any new queen that's introduced after the fact will likely be killed. Unfortunately, it's only a matter of time before the colony dies out.

Under these circumstances, the only way to save the colony is to suppress the ovary functioning of laying workers, which is done by adding open worker brood to the hive.

Introducing Open Worker Brood

INSTRUCTIONS

1. While eggs take three days to hatch, larvae remain "open" for five and a half to six days. Since the larvae produce the pheromones needed, add a new frame of open brood every five to six days when the larvae are young. Do so more frequently with older larvae.

2. As soon as supersedure cells begin to appear, it's a sign that the bees are ready to accept a new queen. You can then remove the supersedure cells and introduce a caged queen.

When a laying worker colony consists of many healthy bees, you may want to try saving it, but take into account that whenever you remove brood from a queenright colony, it gets weaker. Is it worth the risk? If a colony has been queenless for a while, the remaining laying workers may be too small in number and overly aggressive to be worth saving.

For an alternative to using open brood, there is another solution to fixing a laying worker problem, which has worked many times for me and is now my go-to method. As you'll see, once laying workers are shaken out of the hive, they tend not to fly immediately, but rather walk. They seem to love marching from the ground back into the hive. For some reason (which I can't explain), entering the new hive setup suppresses their need to lay eggs.

Removing & Reintroducing Laying Workers to a Hive

MATERIALS/EQUIPMENT

- 1 frame of fertile brood from a queenright colony
- virgin or mated queen, caged
- fresh frames with drawn comb or new frames and foundation
- wooden plank

INSTRUCTIONS

1. Dismantle the hive entirely, shaking the bees from each frame onto the ground in front of your hive. Place these frames aside, as they will still contain laying worker pheromones.
2. Now that the brood box is empty, add a single frame of fertile brood, along with a caged virgin or mated queen.
3. Before closing the box, fill the remainder of the space with new frames and foundation if drawn comb is unavailable.
4. Next, place a board between the ground and hive entrance for your bees to march back into the hive. (Surprisingly, they prefer walking over flying!)
5. It may take several hours for your bees to fully re-occupy the brood box. During this time, place the laying worker frames in the freezer for a period of 24 hours, which will kill the brood. You should also periodically observe the new queen and monitor how your bees react to her.
6. The following day, the frames that were used as space fillers can be replaced by the frames of frozen laying worker brood, which the bees will naturally clean out.
7. After 14 days or so, look for the sign of eggs, which will indicate that you have successfully fixed your laying worker problem.

Introducing a virgin queen to a colony with laying worker bees can be advantageous because virgin queens emit pheromones distinct from mated queens. Laying workers may view a mated queen as a threat to their reproductive dominance. A virgin queen's pheromones can help override the inhibitory effects of laying worker bee pheromones, increasing the likelihood of acceptance.

**Watch how to take care of a
laying worker colony in this video:**
https://beekeep.blue/Svn

Splitting A Colony

Splitting a colony before swarming happens in the spring is a smart move. Believe me, it's much simpler to make a split than it is to catch a swarm! You'll know it's time when worker bees start building queen cells.

Splitting involves taking part of an established colony and rehoming it in a separate hive with plenty of worker bees and honey/pollen stores, as well as its own queen. The bees that stay with the old queen become the "parent" colony, and the ones that go with the new queen become the "daughter" colony.

IMPORTANT CONSIDERATIONS

- Wait until after your first year of beekeeping to split your colony, as the process requires a strong and healthy hive with increased numbers.
- If a split is too small, it won't make it. At a minimum, it should include three combs of brood and one frame of honey with a good number of nurse bees. They are usually found on frames with open or uncapped brood.
- Spring is the best time to make splits because your bees will have plenty of foraging opportunities and time to recover their numbers and stocks. Remember to feed your bees sugar syrup during this period to ensure they have plenty to eat until fresh nectar is available.
- Closely monitor the newly split hives for about three weeks to see if adding another brood frame is required. Since most bees only live about five to six weeks in warm weather, introducing an additional brood frame can support population recovery.
- When feeding, reduce the entrance on smaller colonies to lower the chances of robbing. For splits that are more heavily populated, install a robbing screen.

1.

2.

3.

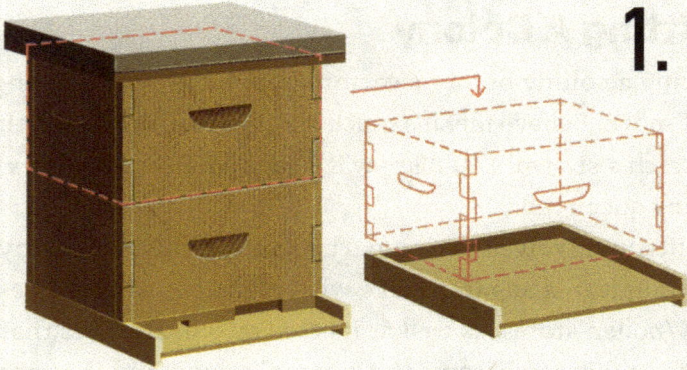

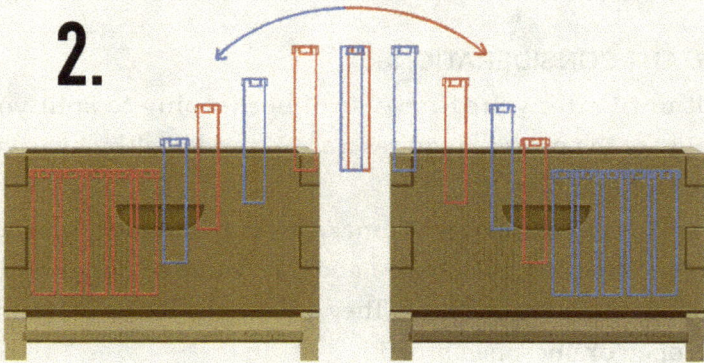

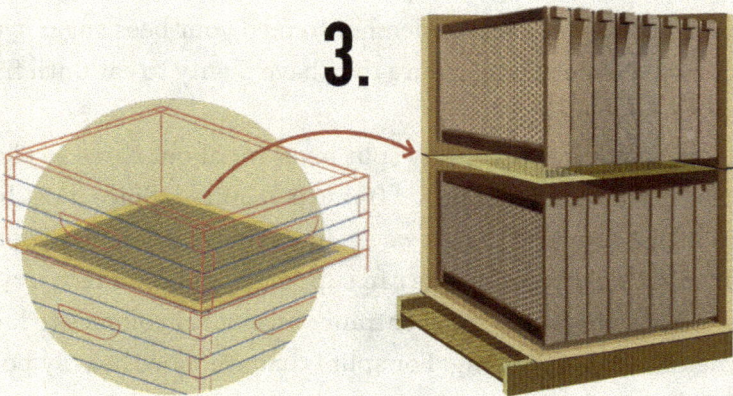

Splitting a colony using a queen excluder

There are several ways to successfully make a split. Here are two of the easiest methods:

A) Splitting a colony using a queen excluder

This method works with a strong two-story hive and should be carried out during the daytime.

MATERIALS/EQUIPMENT

- extra bottom board
- extra lid
- queen excluder
- queen bee (arrange to have her arrive four days after you've added the queen excluder)
- temporary apiary at least three miles away

INSTRUCTIONS

Four days before the scheduled arrival of the new queen:

1. Place the extra bottom board next to the two-story parent colony. Place the top box on the bottom board so you can work with the boxes side by side.
2. Divide frames evenly between the two boxes. Rearrange them so each box contains brood, worker bees, honey and pollen.
3. Place the queen excluder between the boxes and reassemble the hive stack.

Three days later:

4. Look for eggs. Since they take three days to hatch, you'll know the box with eggs is the one that contains the queen. While she continues to lay, the other box remains queenless.
5. Move the queenless colony to a temporary apiary three miles away and leave it there for another seven days. Once the colonies are separated, the queen's pheromones will slowly dissipate.

1.

2.

3.

parent   daughter

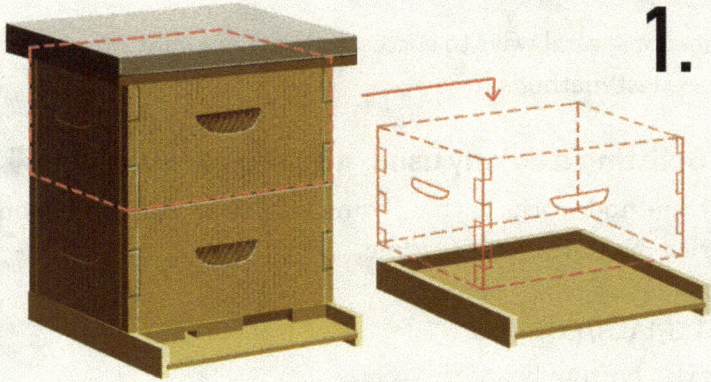

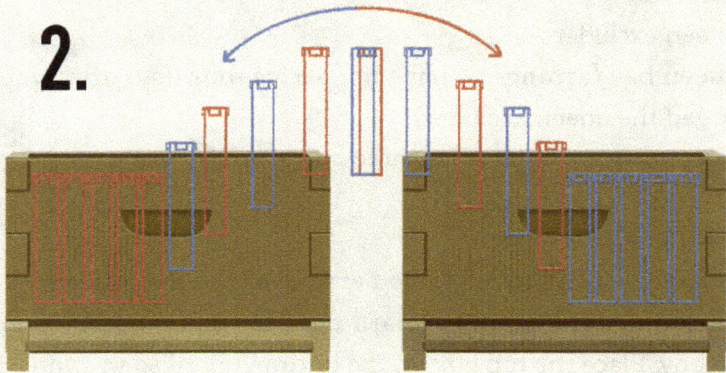

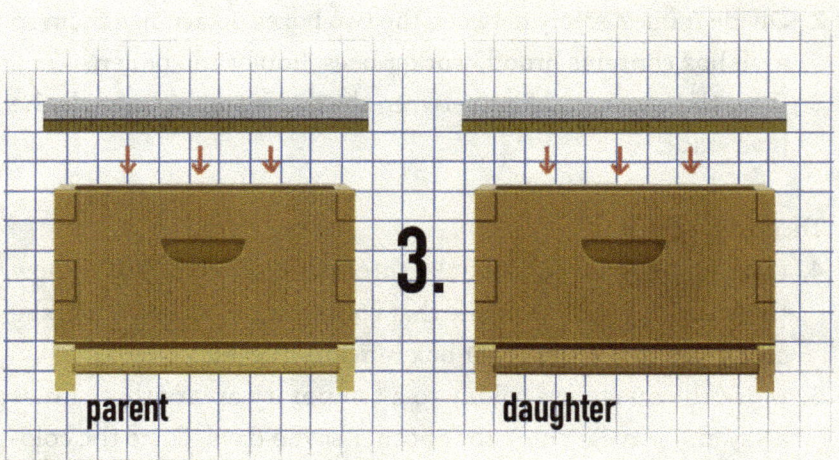

Splitting a colony without a queen excluder

The queenless colony will begin to sense this absence, making the acceptance of a new queen much easier.

Then, when your new queen arrives:

6. Install the new queen in the daughter colony following the steps in the "Introducing a New Queen" section below.

7. After seven days, the daughter colony can be returned to your bee yard. To prevent bees from hive drifting, you may want to implement one of the measures described in the following section.

If a new queen is placed into a colony with an existing queen, let's just say that it won't end well (more grimly, the new queen will be killed by the worker bees).

8. **Seven to ten days later,** inspect each colony, checking for eggs and larvae–seeing these signs in both colonies are an indication that the split has been successful.

B) Splitting a colony without a queen excluder

Dividing a colony without a queen excluder requires the same elements as the first method but saves you from buying and having to store this additional piece of equipment. You'll follow very similar steps to those mentioned above.

MATERIALS/EQUIPMENT

- extra bottom board
- extra lid
- queen bee (arrange to have her arrive four days after you've added the queen excluder)
- temporary apiary at least three miles away

INSTRUCTIONS

Four days before the scheduled arrival of the new queen:

1. Place the extra bottom board next to the two-story parent colony. Place the top box on top of the bottom board.
2. Split frames evenly between the two boxes. Rearrange them so each box contains brood, worker bees, honey and pollen.
3. Place a lid on each box. You now have a parent and a daughter colony.
4. If feasible, move one of the colonies three miles away to a temporary apiary to prevent hive drifting. Alternative methods are also included in the following section.

Three days later:

5. Check for eggs in each of the boxes. Remember—whichever box shows sign of eggs contains the queen bee.

When the new queen has arrived:

6. Introduce the new queen to the queenless colony, as outlined in the "Introducing a New Queen" section below.
7. **Five to seven days later**, inspect both colonies for eggs and larvae.
8. **Seven days later**, the colony placed in the temporary apiary can be returned to your bee yard. Prevent hive drift by implementing one of the measures described below.

By the summer, your parent colony will produce honey while the daughter colony continues to mature and establish itself. Given the average lifespan of a hive, your parent colony probably won't survive a second winter without having to replace the queen, making the daughter colony crucial to keep your bees going in the third year. While regularly checking in on the queen's health, be sure to re-queen as soon as you notice her performance slipping.

Addressing Hive Drift

When making splits, forager bees may return to the original hive's location in the same bee yard. This is called **hive drift**. When this happens, the newer hive loses bees, which limits its chances of survival.

Aside from keeping your splits three miles apart, other ways to prevent bees from hive drifting include:

- Introducing enough nurse bees into the new hive. Nurse bees don't tend to leave their hive, so there's no risk of them flying back to the original one.
- Placing both hives next to each other so traffic is divided.
- Restricting the hive entrance with tree branches (it must be a thick barrier). The barrier acts as a deterrent that forces bees to re-orient before flying off. This trick has personally been very helpful in my own beekeeping practice.

Introducing a New Queen

Once measures for preventing hive drift have been put into place, you will either introduce a caged queen to the daughter colony or allow your bees to raise a new queen from the eggs previously laid by the original queen.. A colony is considered **queenright** when the newly emerged or introduced queen has mated and has started laying eggs. Until then, she is only a potential queen.

Not All Queens Are Created Equal

While allowing bees to raise their own queen can save money and strengthen the genes of the original colony, this option is more of a gamble because not all colonies will have the capacity to raise a healthy queen. Some queens may not return from their mating flights, and others may have poor laying patterns, which won't provide enough eggs to sustain the colony's growth. If you live in an Africanized honeybee zone, allowing your bees to raise their own queen can make them more defensive.

Since it takes three to five weeks for a new queen to fully develop, mate and start laying eggs, opting for a caged queen saves time. In a situation where queen replacement is urgent (i.e., if the colony is suddenly left queenless and hasn't raised a new queen within 16 days), it's best to source a newly mated queen rather than a virgin queen. Mated queens are more likely to be accepted into a new colony having undergone the mating process and having developed behavioral adaptations that enhance their chances of survival: they are more adept at navigating the hive, communicating with worker bees and responding to colony needs compared to virgin queens, which have not yet mated. With a mated queen, the split colony can quickly establish itself and begin building up its population and resources—particularly important during the critical early stages of colony development, as a strong and rapidly growing colony is better able to defend against pests and diseases, forage effectively and produce surplus honey.

When I bought replacement bees in my second year of beekeeping, the queen turned out to be a poor layer. I called the supplier, and luckily, he replaced her for free.

Succession

The demand for queen bees can be high in the spring when most beekeepers split their colonies. It's a good idea to preorder your queen to get ahead of the game. The advantage of opting for a new queen is that she can introduce new genes to the colony that increase resilience and gradually eliminate undesirable traits.

Upon receiving a new queen bee, she'll be in a small cage. Check that she's active and healthy. Store the queen in a warm, dark place until she is ready to be introduced to the hive.

Before replacing a sick, injured or failing queen, she must be removed from the hive first. Otherwise, with the colony on her side, the existing queen will likely attack and try to kill the new queen.

Once you have spotted the existing queen, use a **queen catcher** and remove her at least a few meters away from the hive so her pheromones can no longer be detected. The most humane way of getting rid of her is to place her in the freezer, where she'll fall asleep and die peacefully (*cue Chopin's Funeral March... I know, it's unfortunate*).

After allowing 24 hours for the original queen's pheromones to dissipate from the hive, the colony will then realize that it is queenless. With a more aggressive colony, you might want to extend this waiting period to 48 hours, as it generally takes longer to acknowledge the need for a new matriarch.

Some beekeepers report that if a colony is left queenless for longer than 24 hours, it increases the chances of the new queen being rejected, as the colony may have already started building emergency queen cells around selected larvae.

Inserting the Queen Cage

Most queen cage designs have two release methods: slow and quick. The slow-release method is preferred since the two or three-day delay helps reduce aggressive behavior and increases the likelihood that the new queen is accepted.

You would only resort to the quick-release method if the slow-release approach has somehow failed and the queen has been in the hive for over a week. At this point, the colony will have already familiarized itself with the new queen's pheromones, favoring her acceptance.

There are four ways to insert a queen cage. For each method, the queen cage is placed close to the center of the busiest brood frames. For inserting a queen cage using methods 1–3, find where most of the brood is located within the brood box. If there's more than one in your hive, aim for the bottom box.

I experimented with this once by sticking a cage on the outer edge of the colony. The queen ended up starving because the bees paid no attention to her. Ensuring the queen emerges straight into the busiest part of the hive is your best bet.

METHOD 1

Ensure the exit area outside the queen cage is clear. When you've located where most of the brood is, look for a slight indent in the brood box, indicating that the comb hasn't yet been drawn out.

1. Use your hive tool to open a gap between these two frames, giving you enough space to insert the cage.
2. Next, place the grill of the queen cage against the frame's indent. Check that it isn't touching any wax comb so worker bees can have access to it. This placement allows the colony to interact with and feed the new queen while her pheromones flow through the grill and into the hive.
3. Angle the queen cage at 45 degrees with the candy end facing upwards, ensuring that as the queen eats her way up to escape, nothing can fall in and block her exit.
4. Do your best to push the cage below the top of the frames so it doesn't obstruct the above super or inner cover that will be placed over it. Once the cage is installed, close the hive and leave it for **at least four days**.
5. After this period, check on your queen by looking down into the exit hole of her cage to see if she's eaten her way free. If she has,

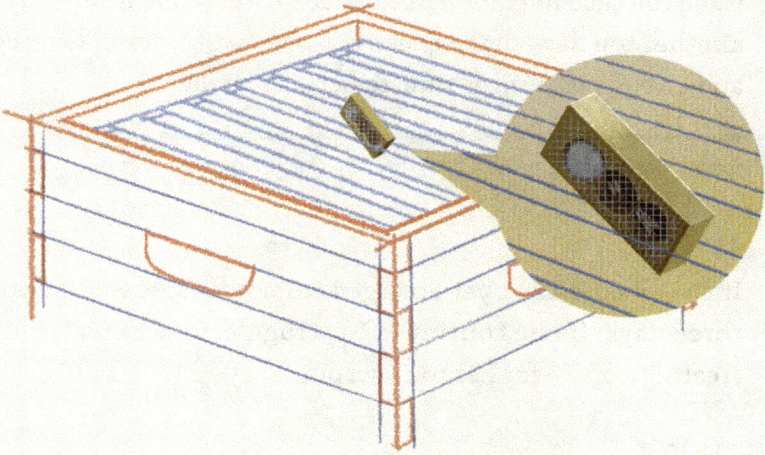

Method 1: *Angle the queen cage at 45 degrees
with the candy end facing upwards*

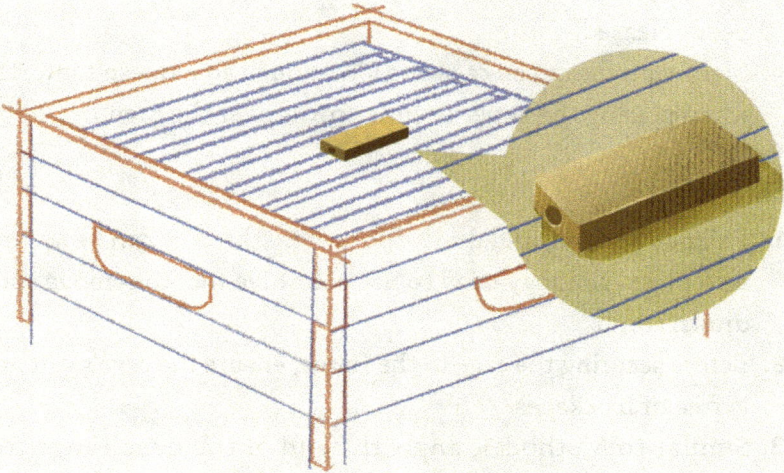

Method 2: *Place the queen cage mesh face down
on top of the busiest brood frames*

leave the cage in place and close the hive. Leave it for **at least another ten days** so the queen can acclimatize, start laying eggs and give the early eggs time to become larvae.

In cooler weather, which can affect egg-laying, give the new queen a little more than ten days to settle in and start laying eggs.

If the queen hasn't yet emerged from her cage, wait another three days. If she continues to struggle, give her a hand by creating a space for her to crawl out.

METHOD 2

1. Before inserting the queen cage, make sure the exit hole is free of any blockages.
2. Place the queen cage mesh face down on top of the busiest, most densely-filled brood frames.
3. Point the end of the cage that contains candy toward the front of the hive, which makes it easier to see whether the queen has been released.
4. Make sure the queen cage sits low enough so it doesn't interfere when placing the super or inner cover over the cage.

METHOD 3

1. Pull out a frame, and find a spot where the cage can be secured into place. You may need to use your hive tool to remove some drawn comb.
2. Before securing the cage to the frame, ensure the cage's exit area is free of blockages.
3. Similar to Method 1, angle the end of the queen cage that contains candy slightly upwards to avoid anything (like dead bees) from falling and blocking the exit.

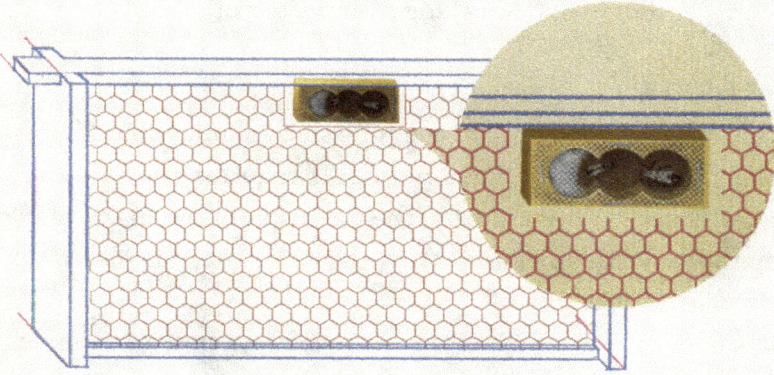

Method 3: *Secure the queen cage directly onto a frame*

Method 4: *An introduction cage can be used with more aggressive colonies*

4. Once the cage is secured, lower the frame back into place, doing so gently to avoid knocking the cage out of place.

METHOD 4

This method is reserved for aggressive colonies that have been queenless for a prolonged period. It requires an introduction cage, which is different from the queen cage used in the previous methods. With an introduction cage, the queen is placed inside after installation.

1. Choose a frame with capped brood that is close to emerging.
2. Gently brush off any bees and slowly press the queen cage into the frame over the capped brood, taking care not to cause any damage.
3. Place the queen inside the cage and seal the entry hole using a candy/fondant plug. Newly emerging worker bees won't realize that the queen has been newly introduced. They will immediately accept the queen and begin feeding her.
4. Acceptance from the newly emerged worker bees gives the queen the green light to enter the rest of the hive, where other bees will most likely follow suit. A few days later, the queen should eventually be released once workers have eaten through the candy plug or tunneled under the cage.
   If this doesn't happen, remove the cage yourself about a week later.
5. Once Her Majesty is released into the hive, remove the introduction cage from the frame.

Your Opinion Counts!

Beekeeping changed my life, and I wrote Beekeeping Blueprint so others could benefit from the same sense of joy and fulfillment that bees offer.

Since most folks pick books based on what others say about them, your review can add to the breadcrumb trail leading others into our wonderful world of beekeeping. To reach as many folks as we can, your support is invaluable! I promise it won't cost you anything more than a short moment of your time, and in turn, it could help:

- **One more family** enjoy the sweetness of their own honey.
- **One more community** discover the importance of pollinators.
- **One more person** find solace and healing in beekeeping.
- **One more garden** become a sanctuary for bees.

**To leave your review,
just scan the QR code or visit:**
https://beekeep.blue/review

If the idea of sharing this wonderful practice with others excites you, then you're just the kind of person I hoped would read this book. Thank you from the bottom of my heart—your recommendation is the gift that keeps on giving.

Let's get back to the bees now, shall we?

Your fellow beekeeper,

Jason Chrisman

"When I was 8, I used to play with this spinning thingamajig in my grandparents' basement. Now I know it was a honey extractor. Many years ago, I developed an interest in beekeeping after going to events where beekeepers spoke about their experiences. I'm 70 now, and I've gone from one nuc to 13 colonies in the last three years."

—David Lewis

CHAPTER 7

A Guide To Harvesting

Going through all the motions of equipping yourself with the right tools, choosing your hive, sourcing your bees and building up your colonies takes hard work. And with hard work comes rewards—let's talk harvesting!

For me, reaching this stage of beekeeping was challenging since I failed to appropriately manage my colonies in my first three years. I started over multiple times, and it wasn't until my fourth year that I could finally harvest my first honey crop.

Before that, I dabbled in pollination services for a local buckwheat farmer after he realized that having beehives around his fields gave him a larger crop. Along with a cousin of mine, I helped pollinate the farmer's crops for two years until every other beekeeper in the county started bringing him hives. I saw a lot of lousy beekeeping habits, and I didn't want my bees to get sick because of it, so I moved on to other beekeeping projects.

Offering pollination services is one example of the many lesser-known ways to make money from bees. While many beekeepers only tend to harvest honey, expanding your range to other products can open up opportunities to engage with your colonies on a different level, unlocking additional revenue streams. As you read through this guide and learn about the wealth of possibilities, I'll also guide you on how to harvest honey, beeswax and pollen, as they're the most common products accessible to a beginner beekeeper.

Earning Money From Your Bees

Deriving an income from your bees might not have been your objective when first setting up your hives, but why not consider it to offset your costs? When you see how much money your bees can bring in with minimal additional effort, the motivation to take good care of your hives is sure to grow. Consider the following possibilities:

SELLING BEE PRODUCTS
- honey
- pollen
- propolis
- royal jelly
- bee venom
- beeswax, from which you can also make/sell beeswax candles

OFFERING POLLINATION SERVICES
As mentioned earlier, this was how I first started generating income from my bees. Renting out your colonies to pollinate farms and horticultural businesses is a simple way to get your bees to pay the bills. Some beekeepers make a full-time income solely from pollination.

PRODUCING BEEKEEPING EQUIPMENT

If you're good at DIY and enjoy making your own equipment, you can start a side hustle by selling to other beekeepers. With original hives, there are plenty of products that aren't standard. These include slatted racks, bee feeders, entrance reducers, bee escapes, etc. If you're good at what you do, you may find yourself building a solid demand for your specially made equipment.

PROVIDING EDUCATION AND CONSULTING SERVICES

The first few years of beekeeping are a learning curve. As you grow in experience, sharing what you've learned with other beekeepers for a reasonable fee is a respectable offering. You could set up classes and demonstrations during the evenings/weekends or provide 1:1 support.

OFFERING SWARM REMOVAL SERVICES

Most beekeepers don't charge for swarm removal unless the bees have moved into a structure. Simply acquiring the bees is the beekeeper's reward.

Swarms are usually easy to relocate, whereas disassembling the wall of a house isn't. A good-sized swarm can be sold in a package for $150 or more. For removing a colony from inside a structure, rates range anywhere from $150 to $1500, depending on the complexity of the situation.

MAINTAINING OTHER PEOPLE'S HIVES

Even with the best intentions, not all beekeepers have the time and energy to care for their bees. For example, horticulturalists or farmers often have beehives to support the pollination of their crops but simply don't have the additional resources to tend to their bees. Once you've gained experience, you can offer hive maintenance services for a fee.

SELLING BEES

Splitting your colony is a great way to make a side income with minimal effort. You can sell your splits if you're not interested in growing your bee yard. In my experience, I quickly learned that selling bees was less work than harvesting honey, so I ran with it.

If you choose to sell bees, make smaller splits or nucs. This way, the existing colony isn't overly impacted, and multiple nucs can be created.

Once you've acquired the skills, experience and adequate supply of bees to sell, it can turn into a very profitable business. At the time of writing, a bee package retails at around $125. A five-frame nuc with a queen goes for about $150–$225, and an established hive costs roughly $250–$350.

Harvesting Honey

True honey has many flavors depending on when and where it was harvested. As different nectar sources become available throughout the season, flavors change.

Before harvesting your honey, knowing when nectar flows are expected in your area is essential. The Honeybee Forage Map by the Goddard Space Flight Center is an online resource that could be helpful.[10] Over time, you'll be able to spot the signs in nature yourself.

When nectar is in flow, add supers to your hive one after the other so your bees can store as much as possible. You don't want to harvest too much honey at once, or your hive won't have enough to sustain itself. As a general rule, leave 40–60 lbs of honey in the hive to help your bees through winter.

10   https://honeybeenet.gsfc.nasa.gov/Honeybees/Forage.htm

Honey flavors change depending on their nectar source

It's also important to get your timing right. Harvest too early and you won't get as much honey as you could. Harvest too late and you risk taking too much as temperatures drop, leaving the colony vulnerable in winter. When in doubt, err on the side of caution and harvest only when 75% or more of the honeycomb cells are capped. Use the following steps to guide you:

MATERIALS/EQUIPMENT

- protective clothing, including gloves and veil
- smoker
- hive tool
- bee brush or escape board
- empty super
- uncapping knife, fork, or scratcher
- honey extractor and spigot
- strainer or cheesecloth
- sterilized bottles or jars

INSTRUCTIONS

1. When harvesting honey, I use an escape board to remove bees from the hive. Place the escape board between the topmost brood box and the honey supers one day before you wish to harvest.

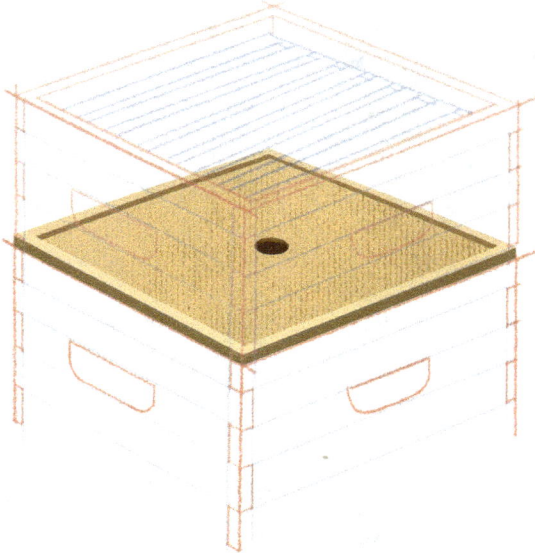

An escape board or bee escape allows bees to exit the honey super but makes it difficult for them to re-enter

2. On harvesting day, put on your protective clothing—as a bare minimum, wear elbow-length gloves and a veil.

3. Open the hive as if you were about to carry out an inspection (Refer to Chapter 5.)

4. Then, if not already done with an escape board, remove your bees from the honey super. Whether you take out a full super or a single frame, ensure it's free of bees. If using a bee brush, gently move your bees out of the way. Avoid any scrubbing motions, as this will only cause harm.

Use gentle flicking motions to clear your bees away

Use an uncapping knife, fork or scratcher to uncap the honeycomb

Once the frames are cleared, set them aside in an empty, lidded super until they're ready to be taken away for honey extraction. If you don't plan on extracting honey right away, it's best to store the frames indoors.

5. When ready for extraction, uncap the wax-sealed honeycomb on both sides of the frame using an uncapping knife, fork or scratcher.

6. Next, place the uncapped frames inside the honey extractor. Spin the frames to force the honey to the sides of the drum, where it will all drip to the bottom.

If using an electric extractor, be careful when powering it up as it can jump around if the weight isn't evenly distributed.

After spinning the frames, the honey is ready to be filtered

7. Place the strainer over the bucket and let the honey flow. The extractor's drum should have a spigot to release the honey. As it leaves the spigot, the honey will filter and drip into the bucket— this is also an excellent opportunity to collect wax cappings. As the bucket fills, close the spigot and replace the bucket. Repeat as needed.

Here's a demonstration of how to uncap, extract and strain honey:
https://beekeep.blue/qbQ

Harvesting Beeswax

Beeswax is produced by special glands in the worker bee's abdomen. As wax droplets are secreted and exposed to air, they become flakes, which are then used to draw honeycomb.

Beeswax has many uses and is most commonly incorporated into skin and hair products. It's one of the top-selling bee products out there. Some people even enjoy eating beeswax for its sweetness and chewy texture.

Harvest your beeswax by collecting all the excess wax found on top of frames or against hive walls during hive inspections. These chunks of extra comb are called burr comb and can get in the way of your beekeeping duties, so removing it just makes sense. Store the collected beeswax in a covered pail or bucket.

Burr comb is the excess or irregular comb built in unwanted areas of the hive, while cross comb specifically refers to the misalignment of comb with the frames or foundation. Both types of comb can cause inconveniences to beekeepers.

Beeswax is a top-seller when it comes to bee products

When it comes to wax cappings, I would advise against pulling frames and scraping from them. Doing that will only leave a sticky mess running everywhere, encouraging hive robbing from other insects. Think of it like standing outside a steakhouse or bakery— the smell is just too enticing! Add that to a lack of nectar flow, and it's a sure recipe for problems.

Instead, as explained earlier, stick to collecting wax cappings when harvesting and filtering your honey. Wax cappings are the easiest type of wax to process because they contain the least pollutants (propolis, debris, pollen). They do, however, contain some honey, which should be removed to prevent wastage.

MATERIALS/EQUIPMENT

- protective clothing
- smoker
- hive tool
- pail or bucket

INSTRUCTIONS

1. Since you'll likely be collecting burr comb while conducting a hive inspection, you should be in your full protective gear.

2. Once you've spotted the excess comb found on top of frames or against hive walls, use your smoker to remove any bees hanging around.

3. Then, use your hive tool to gently scrape off the burr comb and place it in the bucket/pail.

4. Note that burr comb will most likely contain nectar or honey. For an easy way to remove these contents, leave the burr comb out in the open away from the hives, which prevents robbing. Any bees in the area will be able to access the nectar or honey, which they'll then take back to their humble abode.

5. Once emptied, the burr comb is ready to be stored until there's enough to be rendered into blocks. To protect the comb from wax moths, keep in a sealed container at a cool temperature.

Burr comb is the excess or irregular comb built in unwanted areas of the hive

Rendering Beeswax

MATERIALS/EQUIPMENT

- burr comb or wax cappings
- cheesecloth
- clip or twist tie
- saucepan/pot
- water
- tongs
- bucket (optional)
- storage container
- ice cube tray/silicone mold/muffin tin (optional)

INSTRUCTIONS

1. Lay out some cheesecloth and place the burr comb/wax cappings in the middle. Create a bundle and place it in a second layer of cheesecloth.

Using more than one piece of cheesecloth helps reduce the amount of debris getting into your beeswax. If any debris does get through, don't worry. It will be filtered out later on.

2. Secure the top of the cheesecloth bundle with a clip or twist tie.
3. Fill a large saucepan about two-thirds of the way with water. Place the cheesecloth bundle in the water so it's completely submerged.
4. Place the pot on a medium heat setting. As the water warms, the beeswax will begin melting away. As the wax seeps through the cheesecloth, the bundle will shrink.
5. When most of the wax is gone, lift the cloth with your non-dominant hand. Use tongs to squeeze the bundle and extract any remaining beeswax.

Ensure the cheesecloth bundle is completely submerged

It's best to use an old saucepan in the rendering process, as the wax will tend to stick. I've gotten in trouble a few times with my wife for using her good cooking pots, so I know from experience…

6. Once you've extracted as much wax as possible, dispose of the cheesecloth.
7. Turn off the heating element and **allow up to four hours** for the cold wax to cool and form a solid layer on top of the water.
8. Once the wax is solid, hold the pot over the sink or a bucket and gently pour the water out, using one hand to hold the wax in place. If pouring over a sink, drain the water ever so carefully, as wax can easily plug up the drain!
9. Remove the beeswax disk and scrape any debris from the bottom. As needed, repeat the above steps, scraping the bottom of the disk each time to remove any remaining debris.
10. Store the beeswax disk in a sealed container in a cool, dry place. Keep it away from dust and other contaminants. Alternatively, you can melt the wax and pour it into ice cube trays, silicone molds or muffin tins for easy storage.

Extract any remaining beeswax before disposing of the cheesecloth

Wax moths will destroy beeswax. Research also shows that they can eat through plastic bags, which is why storing it in a sealed container is best.

Harvesting Bee Pollen

While honey and wax are the most common products people think of when it comes to bees, you can also harvest bee pollen. Pollen is a good source of protein and contains over 200 nutrients, such as amino acids, essential fatty acids, vitamins and flavonoids. It also has anti-inflammatory properties and may equally help with seasonal allergies.

Pollen is one of the most straightforward products to harvest. Simply attach a pollen trap to the entrance of your hive. As bees

Pollen is a good source of protein and contains over 200 nutrients

enter, the trap gently scrapes away some of the pollen from their legs, collecting it below the hive for easy harvesting. This process encourages your bees to go out and gather even more pollen. It's important to note, however, that pollen traps should not be left on for more than a few days at a time, as bees also require pollen for brood rearing. I'd say a period of two to three days on and off is a safe time interval.

Unlike raw honey, bee pollen can quickly spoil due to bacteria, mold or fungi. Quality control and timing are crucial. If you intend to sell pollen for human consumption, take the necessary precautions to harvest it under hygienic conditions to prevent contamination. Once harvested, process the pollen as soon as possible.

Processing Bee Pollen

There are three stages to processing pollen: drying, cleaning and storage.

MATERIALS/EQUIPMENT

- shallow trays or old window screens
- dehumidifier (if applicable)

- pollen moisture analysis kit (optional)
- small fan
- pollen cleaner (alternative method)
- airtight container

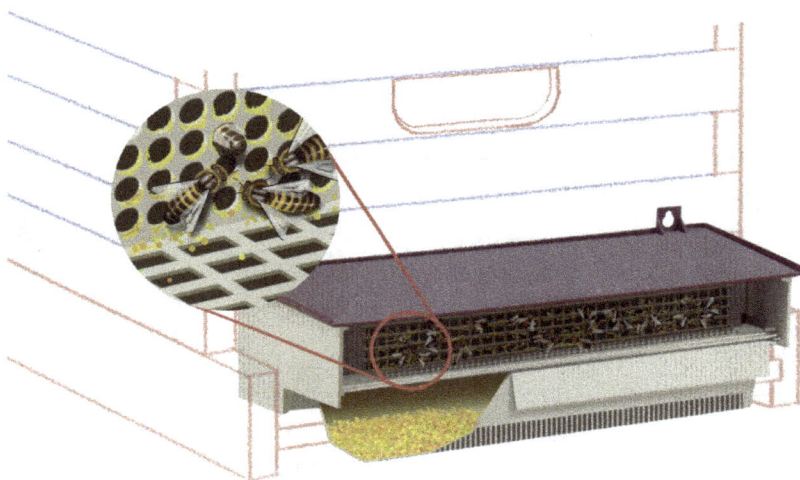

A pollen trap gently scrapes the pollen from your bees' legs as they enter the hive

INSTRUCTIONS

1. **Drying** pollen prevents mold growth and spoilage. Spread the pollen evenly across your shallow trays or old window screens. Leave these trays in a sheltered area away from bees and direct sunlight.

2. Keep the pollen warm and dry at a temperature of around 113°F (45°C). Newly harvested pollen has a moisture content of between 7–21% and for storage, this must be reduced to 2.5–6%. A dehumidifier will be handy if you live in a highly humid area. To check your pollen's moisture content, you can also use an analysis kit and send it to a lab for testing.

If the pollen will be used to feed bees, don't dry it, but freeze it imme-diately after harvesting.

3. Next, it's time for **cleaning**. Use a small fan to create airflow, which will continue to dry the pollen and blow away any foreign particles (bee wings, legs, other bugs, etc.). Anything left over can be picked out by hand.
   Alternatively, a pollen cleaner can be a quick and easy method to remove unwanted debris.
4. Once the pollen has been dried and cleaned, **store it** in a clean airtight container. Pollen is hygroscopic, so it will absorb mois-ture if exposed to the air, affecting its quality. Freeze for 24–48 hours to help avoid insect infestations.

Pollen will have a shelf life of about 12 months, so sell it while it's fresh!

"I was allergic to bees but always wanted to get into beekeeping, so I watched hours of beekeeping videos. I got some lumber and built some swarm traps. I caught 20 the first year. Had a terrible time— the first stings had me blown up like a balloon with my eyes swelled shut for days, but I never gave up. Now they don't affect me at all except for the minor pain. I have about 15 strong hives with a great honey harvest this year. I enjoy every minute with my bees, and some of the best relaxation I get is watching them work."

—Michael Fike

CHAPTER 8

Troubleshooting Pests & Disease

There's a common assumption that ants and bears are the only problems bees have, but in reality, there are many pests and diseases that can give bees a hard time. In the beekeeping world, there are three schools of thought when it comes to pest and disease control: using the latest proven products, using natural treatments (such as essential oils or controlled breaks in the brood cycle), or simply allowing nature to take its course (i.e., the strong bees survive and the weak die off).

Since bees deserve the best care, in my beekeeping practice, I do my best to stay up to speed with the latest treatment products. In this chapter, I'll share some best practices for troubleshooting pests like mites, hive beetles and ants, among others. I'll also go over the most common diseases that can affect bees and what to look out for so you can take action.

At the first sign of pests or disease in your colonies, be sure to supplement the information in this book with your own research for the most appropriate treatment (if any). You may also want to consult a local expert for advice. Occasionally, after exhausting all options, you may have no choice but to dismantle a hive. In these worst-case scenarios, take comfort in knowing that by doing so, you're bringing relief to your bees and alleviating their suffering.

Varroa Destructor: Enemy #1

The Varroa mite is the world's most destructive honeybee pest, causing a significant decline in honey production and pollination services. Scientifically, it's no surprise that this pest is known as Varroa *destructor*.

During my first three years as a beekeeper, I started over and over again because I didn't want to deal with the pest issues that

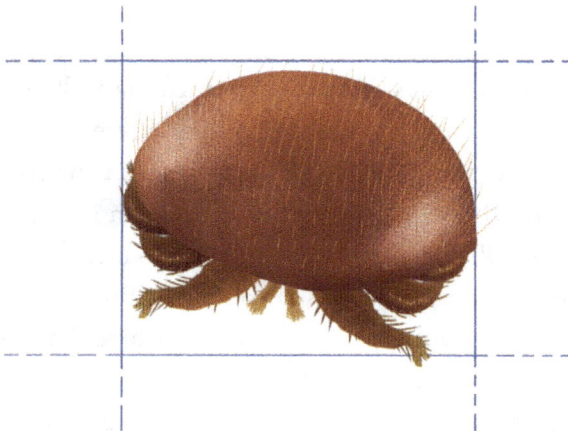

Varroa mites feed on an organ known as the bee's fat body

kept coming up. My bees would die, and I'd simply go out and buy new ones. As I was about to purchase more bees to start over, I questioned whether beekeeping was for me. I was so frustrated! Beekeeping isn't cheap and having to buy new bees every year added to the expense.

Nevertheless, I decided to give it one more go. At the same time, I realized my attitude had to change and the bees needed to be treated like livestock, not just a hobby. While a hobby can be put on hold and picked up again later, livestock becomes deadstock unless tended to.

Once I finally began treating for mites, magical things started to happen. My bees thrived, and the number of active beehives I kept flourished. Before I share how to test for and treat a Varroa mite problem, let's take a closer look at how these pests operate. It's all about keeping your friends close, but your enemies closer.

Reproduction & Feeding

Varroa mites like to congregate on bees working near a hive's open brood. This allows them to quickly enter mature larvae cells for reproduction before the cells are capped. For this reason, young nurse bees tending to brood are usually the most affected by Varroa mites in a colony.

Varroa mites tend to avoid queen cells due to their aversion to the royal jelly fed to queen larvae. Brood cells containing drone and worker bees are preferred, as the post-capping stage for these bees is longer, giving the Varroa mite more time to reproduce.

Varroa enjoys feeding on what's known as the bee's fat body—the organ that provides energy to survive non-foraging periods such as winter. Feeding on the fat body weakens the bee's metabolism and can also transmit viruses, including deformed wing virus or acute bee paralysis virus.

Watch this to learn more about the reproductive cycle of the Varroa mite:
https://beekeep.blue/2mb

Testing for Mites

Mites are about the size of the ball on a ballpoint pen, so it takes a trained eye to spot them. Many new beekeepers say, "I didn't see any mites, so I didn't treat," only to have their hive later ravaged. They either decide to give up beekeeping altogether or put in more money to start all over again, just as I did. Even if you don't see mites, they may be hidden under capped brood cells or clinging to the underside of your bees. Testing your colony is vital.

In each of my hives, I measure the number of mites per 100 bees monthly so I can keep on top of when treatment is necessary. Only a small sample needs testing to determine the overall situation of a hive. There are several ways to test for mites, including doing an alcohol wash, sugar roll test or using sticky boards. Since alcohol washing provides the most accurate results, this is the method I'll share with you.

ALCOHOL WASHING

The only downside of doing an **alcohol wash** is that you'll lose around 300 bees from your hive in the testing process, which many beekeepers prefer to avoid. When we consider that queens produce 1000–1500 eggs daily, however, sacrificing 300 bees for the health of the entire colony doesn't seem so bad.

Ensure the queen is not part of your test sample

When testing, make sure the queen bee isn't among the sample. What I suggest is to locate the frame with the queen so it can be moved to a spare box, ensuring the queen is kept safe.

Before doing an alcohol wash, be ready with the proper gear and the necessary supplies.

MATERIALS/EQUIPMENT
- sampling device (like Varroa EasyCheck or Varroa Mite Sample Bottle)
- ½-cup measuring scoop (optional)
- isopropyl alcohol (75% or higher) or non-foaming winter wind-shield washer fluid
- large plastic tub
- fine mesh strainer
- quart-sized plastic recipient to use during straining
- large, lidded container for storing used sampling fluid (optional)
- access to water

Isopropyl alcohol and flames do not mix, so keep your smoker at a distance.

INSTRUCTIONS

1. Fill the alcohol wash sampling device with the isopropyl alcohol or windshield washer fluid until it covers the bottom of the inner basket or cylinder.

2. Find and isolate the frame where the queen is. Temporarily place this frame in an empty box.

3. Next, select your bee sample for testing. The sample should contain anywhere from 200–300 bees. Choose a frame at the outer edge of the brood area so you can aim for as many nurse bees caring for open brood as possible.

4. There are two ways you can collect bees for your sample:

   i. Hold the selected frame firmly as you give it a good shake over the plastic tub. The bees will drop into the tub. As you lift one corner, tap it firmly to move the bees to one side.
   Use a ½-cup measuring scoop to gather your test sample. Remove the lid of the sampling device and drop the bees into the inner cylinder before replacing the lid.

   ii. Either brush the bees from the frame into the empty cylinder of the sampling device or slide the basket gently down the frame and across the back of the bees—somehow this magically knocks them right into the cup!
   Continue until the bees reach the cylinder's fill line, which amounts to a sample size of about 300. Place the cylinder back into the sampling container and replace the lid.

5. Shake the sampling device a few times to wet the bees with the alcohol/windshield wiper fluid. Open the sampling device once more and top it up with fluid until the container's upper line.

6. After securing the lid back in place, gently shake the container for one minute in all directions: vertically, horizontally and in a circular motion.

7. Holding the container up to the light, count the number of mites that have collected at the bottom. Divide this number by

(i) Shake your bees into the tub, then lift the tub from one corner, tapping it down firmly to move the bees to one side for collection

(ii) Slide the sampling device basket down the frame

three to get a percentage (e.g., 12 mites per 300 bees would give you four mites per 100 bees or 4%).

While some beekeepers will allow for a 3% infestation threshold before treating, experience has taught me to play it safe and **treat with anything over 2%**.

8. Once you've determined whether treatment is necessary, lift the basket from the sampling device and pour the sampling fluid through the strainer and into a recipient to filter out mites and debris. The sampling fluid can then be reused several times for future testing.

9. Before discarding the test sample, I like to re-wash the bees with plain water to confirm that all mites have been dislodged.

Want to see how to perform an alcohol wash? Check this out:
https://beekeep.blue/UQf

Prevention & Treatment

When it comes to Varroa mite control, approaches range from prevention to intervention as you deal with more serious mite problems.

CULTURAL APPROACHES

Used as preventive measures, the following cultural approaches help reduce Varroa mite reproduction. Keep in mind that none of these measures are foolproof:

- **Choose mite-resistant bees** or Varroa Sensitive Hygiene (VSH) bees that can recognize and remove mite-infested brood. Sometimes, however, the VSH gene can make the bees more defensive.

- **Use small cell comb** to mimic comb found in the wild (~4.9 mm in size vs. 5.4 mm in commercial foundation). Although the effectiveness of this approach is debated, the shorter post-capping period in smaller cells is linked to fewer Varroa mites in each cell.
- **Take a brood break** by caging the queen for about three weeks. During this period, mites will be forced out of cells and onto adult bees once the brood hatches. This approach, although effective, requires precise timing at each phase of the process. When combined with chemical treatment, brood breaks can especially help reduce the presence of Varroa in your hive.

Splitting your colony is another option, as this gives the daughter colony a brood break while it prepares to raise a new queen. If you decide to introduce a new queen, be sure to remove any queen cells as they appear.

MECHANICAL APPROACHES

These intervention methods are most effective when used in combination with another:

- **Add a frame of drone comb to your colony** to encourage drone production and lure mites into the brood cells. Before the drones emerge, remove the frame and place it in the freezer, which will kill the larvae. After returning the frame to the colony, the bees will do a natural clean up of the frame. This approach delays mite reproduction and lowers the number of mites in your hive. It may not, however, be enough to eliminate Varroa mites entirely.
- **Sprinkle powdered sugar on your bees.** Although this doesn't eliminate the mite population, it encourages grooming and helps increase mite drop.

- **Use a screened bottom board**. As bees move around the colony and groom themselves, mites will naturally fall to the ground and are unlikely to climb back into the hive.

CHEMICAL APPROACHES

If ever you find yourself with a large mite population in the fall, I strongly recommend using chemical miticides before the colony starts producing winter bees. It will be your colony's best chance of survival during the cold season. When having to resort to a chemical approach, try soft chemicals before opting for harder chemicals that will have more severe repercussions on your hive and bees.

Soft chemicals are made from natural products. These provide effective treatments that don't leave any chemical residue in your hive. They include: formic acid, oxalic acid, thymol and hop beta acids

Hard chemicals will usually get rid of up to 95% of mites. Over time, mites have gradually become resistant to some of the most commonly used chemicals (like fluvalinate and coumaphos). They also leave behind residue that tends to accumulate in beeswax. This residue can harm your bees and make them more susceptible to Nosema disease.

These days, the most prevalent chemical in use is amitraz, which doesn't seem to contaminate honey or wax. Amitraz can be found commercially under the name Apivar.

I hate to break it to you, but as a beekeeper, the fight against mites is long and arduous. Even when we're on top of things, our bees will still meet bees from other colonies when foraging, where they risk exposure to more mites. The most we can do is carry out regular tests, take steps to keep mite numbers down in our hives, cross our fingers and hope for the best.

Small Hive Beetles (SHBs)

In great numbers, small hive beetles can cause significant stress to bees, which can devastate a colony if it's already weak or dealing with other stressors. SHBs lay an extensive number of eggs that develop quickly. They feed on pollen, honey and bee brood, destroying any unprotected comb in a hive.

SHBs are easily spotted with the naked eye. To help eliminate them, there are **traps containing vegetable or mineral oil** that can be used.

While there are also chemical alternatives, I recommend **releasing nematode worms into the ground** to stop SHBs. Mix the nematode worms in water and pour them into the soil using a simple watering can or a pressurized sprayer. As nematodes burrow

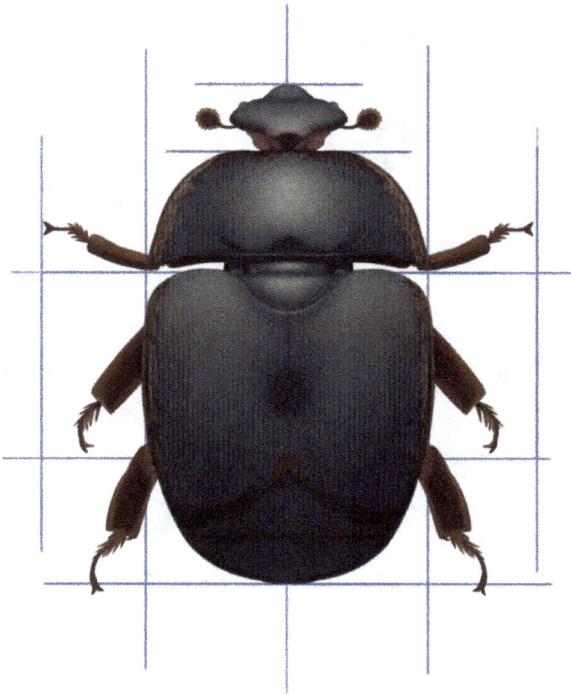

Small hive beetles feed on pollen, honey and bee brood

into the soil, they will target SHB larvae or pupae, entering the insect's body and releasing deadly bacteria.

As they reproduce, nematodes will then spread further out around your apiary. Depending on the soil type, this can have an impact on the effectiveness of this treatment. Nematodes may also not survive winter or periods of drought.

Ultimately, the best way to steer clear of SHBs is through prevention. Always keep your hives clean, healthy and strong by conducting regular inspections. It's also important to keep your bee yard clean. Avoid leaving frames or chunks of wax or propolis lying around, as these will only attract pests.

Other Unfriendly Insects
Ants

Ants love the sweet honey and nectar available in beehives. While having one or two around isn't an issue, too many ants can cause bees to abandon their hive.

The easiest way to deal with ants is to **raise your hive on a platform.** Place each leg in a can or bucket and add a few inches of oil to create a moat. It may sound a little medieval, but it keeps ants out!

While many beekeepers opt for used motor oil in their buckets, this can potentially contaminate the soil when it rains, so I stick to food-grade oils like vegetable oil. Contaminated soil will also affect the flowers your bees forage, which you definitely want to avoid.

Tracheal Mites

As you can tell by their name, these pests live and reproduce in the trachea of European honeybees. Blocking oxygen flow, they eventually kill the adult bees they infest. Tracheal mites can spread throughout the hive from direct contact between bees. Other

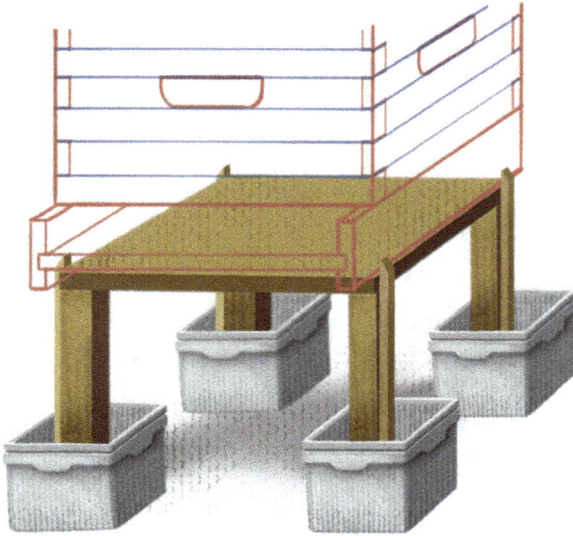

Keep ants out by setting up small moats of food-grade oil at the base of your hive

symptoms include disjointed wings and lethargy—on nice days, you may notice most bees staying in the hive rather than flying.

Take stock of tracheal mites by regularly collecting a sample of 50–75 bees to examine for signs of infection. If necessary, treat with **menthol pellets** (50–60 g in each infected hive). Place the pellets on top of frames when temperatures are below 60°F (~15.5°C) or on the bottom boards when temperatures are above 78°F (~25.5°C). Keep the pellets in your hives for at least two weeks. As the menthol vaporizes, this kills the tracheal mites.

Alternatively, you can make **grease patties** from one part solid vegetable shortening and two to three parts granulated sugar. Shape the patties into hamburger-sized disks and place them on wax paper at the center of your hive's frames. After eating the sugar, your bees will be covered in shortening, which prevents tracheal mites from spreading to other bees. Eventually, they'll die off altogether.

Wasps

In the summer, wasps hunt for sweet, delicious food as their nests reach capacity. When the season draws to a close, their natural food sources dwindle, causing them to invade hives and steal honey. While they're at it, wasps will also attack and bite bees in half, bringing the abdomen back to feed their colony. Isn't that wild?

Wasps are aggressive beehive invaders

As a preventive measure, set up an **entrance reducer** to make it easier for guard bees to protect the hive. You can also leave wasp traps around to catch any scout wasps on the hunt for hives to attack. If you want to avoid attracting wasps, it also helps to keep any honey, wax comb or sugar syrup from spilling near your hive or around your bee yard. When not in use, store brood frames and supers in a secure/protected area.

Bee Lice

Although bee lice have a minor impact on colonies, they can still destroy comb, affect honey production and, in rare cases, infest the queen, impairing her ability to feed. The latter can result in poor egg-laying, affecting a colony's population growth.

Bee lice can find their way to a colony in several ways, including attaching themselves to swarming bees. Drifting bees, packaged or queen bees can also bring lice into the hive. These pests can be found at the bee's mouth, feeding on nectar, pollen and other substances.

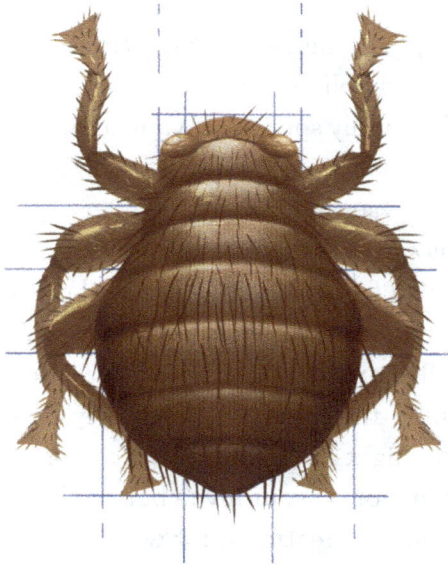

Bee lice feed on nectar and pollen, among other things

Signs to keep an eye out for are burrowing paths in your hive's comb, as bee-louse larvae tunnel across and begin searching for adult bees to infest. You should also be able to physically see bee lice as they're much larger than other mites or pests.

If you suspect bee lice in your hive, **freezing honeycomb** for 48 hours can take care of the problem. You can also place honey immediately in the freezer after extraction—this will destroy any bee-louse eggs or larvae found in the honey. The same approach can be used for other pests as well.

For heavier infestations, place a sticky board on the bottom board of your hive. Fill your smoker with a small amount of tobacco. Smoke the hive moderately, as excessive tobacco-infused smoke can be harmful to bees. After a few minutes, check the sticky board for results.

Greater Wax Moth

These are small, gray-beige moths attracted by the scent of beeswax, honey and pollen. Wax moths often enter the beehive at night and aren't easily spotted. If your colony is strong, worker bees will chase the moths out of the hive. However, bees don't tend to remove moth larvae, and if there aren't enough bees patrolling the comb, moths can eventually take over.

Moth larvae will feed off beeswax, larval cocoon silk and bee feces. Adult moths, on the other hand, love older, darker comb used to store pollen and raise bee brood. Signs to look out for are spider-like webbing and tunnels in the honeycomb left behind by moth larvae. You might also spot larval feces (small black cylinder shapes) in the comb or on your hive's bottom board.

There's no chemical treatment for wax moths that won't hurt your bees, so it's best to have them defend themselves by **keeping your colonies strong and healthy**. As a preventive measure, **avoid placing too many boxes on your hive** if your hive population can't patrol the comb. You can also try setting up moth traps, although these will never be as effective as proper hive management and care.

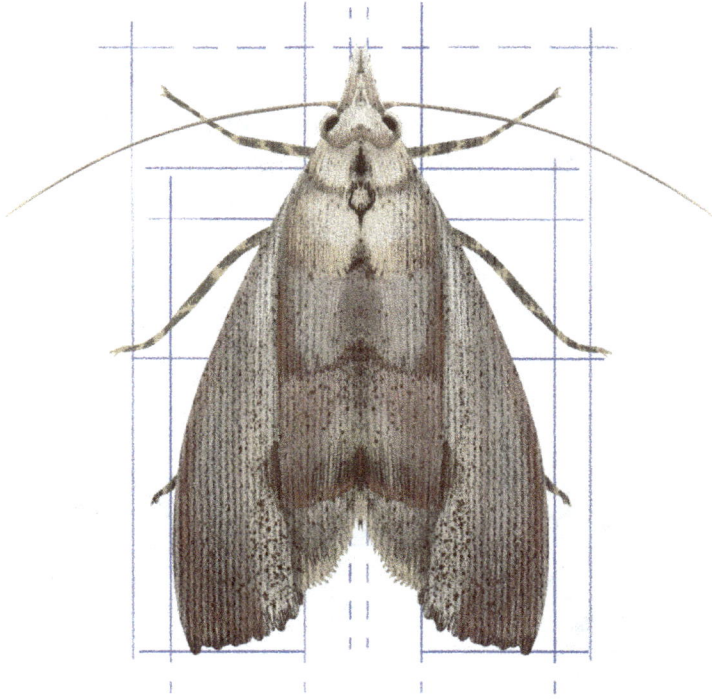

The best way to fend off wax moths is by keeping your colonies strong and healthy

Watch how you can protect your hives from wax moths:
https://beekeep.blue/DWn

Spiders, Earwigs & Cockroaches

Other insects and arachnids can also pose a threat to your bees. Spiders won't hold back from trapping bees in their webs and snacking on them or attacking bees while they're out foraging.

Cockroaches and earwigs, which are often mistaken for one another, are both pests that will move into a hive and deplete food stores. They're usually found on your hive's bottom board and around the inner cover. Don't be afraid to squash these insects with your hive tool! Otherwise, **maintaining strong and healthy colonies** is the most efficient form of pest control.

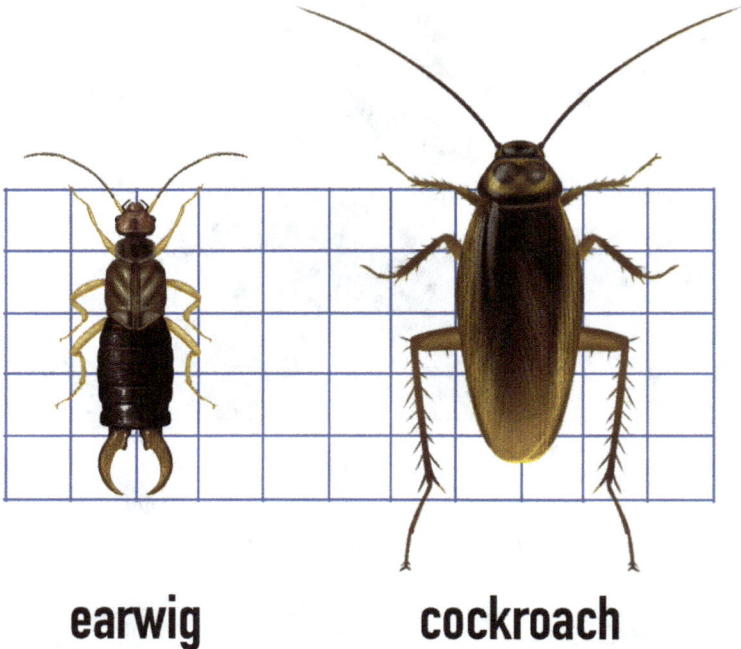

earwig   cockroach

Earwigs and cockroaches can be found on the bottom board or around the inner cover of your hive

Larger Nuisances
Mice

In the winter, mice will happily climb into an occupied hive—not in search of food, but a cozy nest. Although they won't hurt your bees, they'll chew through comb, damaging your hive's foundation and frames. Active bees can generally defend themselves from invading mice, but as soon as temperatures drop and they begin to cluster for warmth, this is when mice move in.

Drawn comb is a huge resource come spring, so you should do everything you can to protect it until then. Keep mice at bay by **reducing the entrance to your hive** to less than eight or nine millimeters. **Installing a flat metal mouse guard** is also an option.

Skunks

As insect eaters, skunks love to feed on bees. These clever animals will scratch at a hive's entrance and snatch up any bees within reach, sucking out their juicy insides and spitting out their exoskeleton—the poor things! When you notice scratches on the lower part of your hive and find bits and pieces of your bees on the ground outside the hive, you've got a skunk problem.

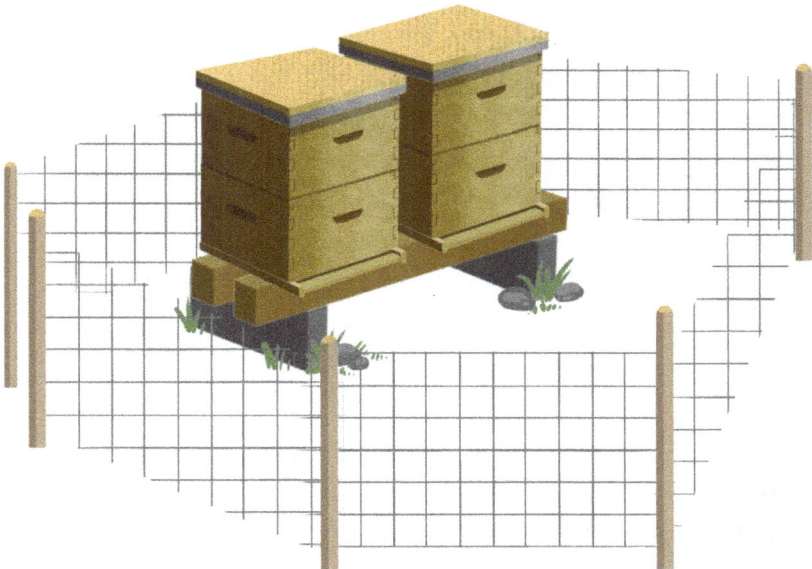

Consider electric netting to keep skunks, racoons and weasels out

My go-to approach for deterring skunks is **placing hives on a stand** several feet above the ground. This forces any tempted skunks to stand on their hind legs to reach for the bees, exposing their sensitive underbelly. Once the bees have identified the threat, they'll target this area and sting. I must say that the added height also does wonders for my back!

Another solution is to **place poultry netting around your hive**. While skunks can't climb this, pests like raccoons and weasels can, so you might want to electrify the netting to keep them away.

As for getting under the netting, skunks may still manage, so **consider setting up a bed of nails or spike stripes.** To do this, pound a few nails into a piece of plywood slightly wider than the hive and place it under the entrance with the nails pointing up—just be careful when moving around your hive.

Bears

Anyone who's seen or read Winnie the Pooh knows how much bears adore honey. Unfortunately, as cute and harmless as Winnie the Pooh may be, any real bear trying to get to honey will destroy your hives. Once a bear has developed a taste for honey and bee brood, it'll keep coming back.

Setting up an electric fence around your hives has a nearly 100% success rate at keeping bears away. Generally, if you live in an area with bears, make sure to site your hives at least 300 feet from any surrounding forest.

Diseases

Honeybees can develop a range of diseases, some of which are highly contagious. At the first sign of disease, take immediate action. Since this book is intended as an introductory guide to beekeeping, we won't cover all the possibilities of disease and their respective treatments, but I will share an overview of the main diseases to watch out for.

For more information on disease prevention and control, check out Main bee diseases: Good beekeeping practices *by UN's Food and Agriculture Organization (FAO).*[11]

Viral Diseases

When bees are affected by a virus, few treatments exist to save them. If the infection is still in its beginning stages, replacing the queen and infected honeycomb could potentially solve the problem. Unfortunately, if it's a particularly severe infection, the only option is to destroy the affected colony.

CHRONIC BEE PARALYSIS VIRUS (CBPV)

Most often found in colonies affected by Varroa mites, bees suffering from CBPV lose most of their hair, turn dark in color and experience nibbling attacks by their healthier counterparts. In the upper part of the honeycomb, you're likely to notice wobbly bees that have trouble flying. You might also find them crawling on the ground or on stems of grass, where they eventually die. With CBPV, affected brood cells become hollow and larvae turn yellowish-brown as their internal organs dissolve. The larvae will then dry out, mummifying into little black flakes.

ACUTE BEE PARALYSIS VIRUS (ABPV)

Alongside the Varroa mite, an ABPV infection can become serious, killing both brood and adult bees. The virus is normally found in the bee's fatty tissue and does not present any symptoms, although is very quick to develop and wipe out a colony.

---

11 https://www.fao.org/3/I9466EN/i9466en.pdf

CBPV affects both brood and adult bees

It's easy to spot a DWV infection

DEFORMED WING VIRUS (DWV)

DWV is commonly found in apiaries, usually triggered in combination with a Varroa mite infestation. When active, DWV will attack developing bees in their cells. Unlike ABPV, this virus replicates slowly. When affected by DWV, bees are still able to fly, although will have a reduced body size, severe wing deformations and a short life expectancy.

BLACK QUEEN CELL VIRUS (BQCV)

As the name suggests, BQCV affects only queen bee cells, turning larvae and cell walls black. It's one of the most common causes of death in queen larvae and tends to strike when a colony is already affected by nosemosis (explained in more detail in the following section). While worker bees and drone brood can also contract this virus, they remain asymptomatic.

SACBROOD VIRUS (SBV)

SBV symptoms are only noticeable in capped larvae cells; adult bees are generally asymptomatic. When infected, capped larvae will turn a yellowish-brown color before their internal organs slowly dissolve. Luckily, this virus is easily taken care of. Subjecting infected frames to temperatures between 131–149 °F (55–65 °C) for 10 minutes or exposing infected frames for six days to direct sunlight can help reduce SBV contamination. Keep in mind that these treatment methods should be used cautiously and in conjunction with other disease management practices.

Fungal Diseases

NOSEMOSIS

Nosemosis, more commonly known as Nosema, is a fungal disease that affects adult bees, although rarely affects the queen. There are two types of this disease: Nosema apis and Nosema ceranae.

Both are usually hard to identify before it's quite advanced. At this point, the beekeeper's only option is quarantine and destruction of the affected hive.

- **Nosema Apis (N. apis):** More common in cold and wet areas, N. apis usually occurs alongside a decrease in a colony's population. Symptoms include intestinal disorders such as diarrhea. N. apis also affects the bee's ability to produce royal jelly, while forager bees become increasingly less active. In the rare case of a queen infection, her egg-laying will decrease significantly.

  Slowly, the affected colony's population dwindles. Some bees will lose their ability to fly and may even become paralyzed. You may also notice dead bees at the bottom of your hive with swollen abdomens and their legs drawn back towards the chest. The most obvious sign of an N. apis infection is the liquid excrement spread across the hive entrance and honeycomb.

- **Nosema Cerenae (N. cerenae):** This nosemosis strain differs from N. apis in that it doesn't affect the intestines. What N. cerenae does is cause forager bees to leave the hive to die, which means you won't notice any casualties as you would with an N. apis infection. A silent killer, N. cerenae leads to a colony's gradual depopulation until it's wiped out entirely.

AMOEBIASIS

Although not a fungal disease, amoebiasis and nosemosis often work hand in hand. Affecting adult bees, the symptoms of this disease are similar to those caused by N. apis, which include a swollen abdomen, quivering wings, problems flying and diarrhea. Amoebiasis can be controlled by removing infected comb, providing the hive with fortified supplements and disinfecting your gear with bleach between each use.

CHALKBROOD

As a fungal disease affecting bee brood, larvae often die within the first two days of capping. Chalkbrood causes larvae to mummify as they dry out and harden. They can turn white, gray or black. If your colony has a chalkbrood infection, you may notice little stones (chalkbrood) at the bottom or entrance of your hive. Eradication is impossible, but a hive will generally pull through with good management practices (i.e., ensuring your bees have enough food and that humidity is kept out of the hive). Some beekeepers might also feed their bees sucrose syrup acidified with lemon juice.

Chalkbrood mummifies bee larvae, turning them into white, gray or black-looking stones

STONEBROOD

Stonebrood is a fungal disease that affects adult bees and both capped/uncapped brood. Initially, larvae will take on a white and fluffy look before turning yellow or greenish-brown as they mummify. Adult bees, on the other hand, will become agitated, weak or paralyzed, losing their ability to fly. You may also notice a swollen abdomen and eventual mummification.

Bacterial Diseases

AMERICAN FOULBROOD (AFB)

AFB is one of the most common, destructive bee diseases out there, causing the irregular capping of honeycomb and high brood mortality rates. With an AFB infection, capped cells become darker, sunken or hollow at the center. You might also notice an unpleasant, sour smell, as larvae turn yellowish and then dark brown, with their bodies finally liquifying into a stringy consistency.

*An AFB infection results in larvae liquifying
into a smelly, stringy consistency*

EUROPEAN FOULBROOD (EFB)

EFB is another bacterial disease that attacks bee brood. While some bee varieties have developed a natural resistance to EFB, there can still be severe consequences when infected. Larvae will die within a few days before the cell is capped. They then turn opaque and yellow/yellowish-brown as they decompose and dissolve into a soft brown mass. This mass then dries up to form a dark rust flake.

When keeping bees, allowing nature to take its course without intervening can have devastating financial impacts, not to mention the loss of all the potential benefits that beekeeping brings. The best advice I can impart about protecting your colonies from pests and disease is to be proactive and maintain good hive management. Prevention is usually much easier to deal with than intervention. You may not get it right on the first try, but you should now know what to keep an eye out for during your regular inspections.

"A friend suggested that I needed something to do after my wife passed away. He said working with the bees would be calming. He was right. It took my mind off things. Twenty years later, it still brings me solace..."

—David J. Little

CHAPTER 9

Continuing The Journey

I believe everyone who starts beekeeping has a wish to be successful. It's probably why you bought this book—you wanted a step-by-step guide on how to care for your bees. Everything I share about working with bees through the seasons has come through trial and error. In my first three years, it seemed like something different kept failing in my hives. I put in the work and continuously made changes until my bees started to thrive. I certainly learned the hard way, but those lessons will never be forgotten, especially now that they're documented here in this book!

From covering essential background information about bees and hives to building up your skills as you establish your first colonies, I hope *Beekeeping Blueprint* has and will continue to come in handy as you take off on this incredible journey.

Even over a decade into beekeeping, I'm still learning about my bees and about myself. Managing dozens of colonies on top of running my farm, I am constantly working on finding a balance. When it's time to split my colonies and make nucs to sell, it's also calving season—handling it all on my own is a challenge I face every year.

We never stop learning as beekeepers, and I've come to find that's part of what makes it special. As the cycle continues, I find the very best way to keep progressing in my knowledge of beekeeping is to compare notes with others.

Moving forward, I hope you'll choose to connect with those who share your passion for beekeeping. As mentioned throughout the book, now is a great time to join a local beekeeping chapter or community group. Online communities and forums are also a wonderful way to share experiences, discuss ideas and solve problems together—the only thing I would caution is to not believe everything you read. As a rule of thumb, make it a habit to verify your information with at least one local beekeeping source. Other than that, have fun experimenting, making mistakes, learning from those mistakes and embracing the process.

I've also founded the *Beekeeping Blueprint Community*, where beekeepers around the world can come together to share their successes, find support, access exclusive content and so much more.

As seen from the personal testimonies featured throughout the book, beekeeping is a practice that keeps on giving. The rewards will last for as long as you decide to keep at it. Beekeeping is much more than a hobby. It's a way of life and a way to connect with yourself through Mother Nature, as well as the wonderful community of other beekeeping enthusiasts out there. *Welcome to the club!*

**Please leave a review to help
others on their beekeeping journey!**
https://beekeep.blue/review

Additional Resources

Some of my favorite beekeeping resources have been mentioned throughout the book. There are a few more, however, that I didn't have room to include and merit special mention:

Randy Oliver's Scientific Beekeeping[12]

Over time, as Varroa mites have become more resistant to certain chemicals, Randy Oliver has done a great job keeping beekeepers informed on the latest treatments. His website has a wealth of knowledge that I recommend to anyone who is a fan of data and evidence-based research. Randy provides practical insights into all aspects of beekeeping with a focus on colony health, disease management, and sustainable practices.

Wikipedia's List of Pollen Sources[13]

I'm just going to go ahead and say it—Wikipedia is an excellent source of information! This particular page has an awesome setup, and it's one that I rely on myself. It includes an extensive list of pollen sources organized by plant species and blooming season. What I find the most helpful is seeing the different colors of pollen.

Access my ongoing list of recommended tools, products and resources:
https://beekeep.blue/cA2

---

12   https://scientificbeekeeping.com/

13   https://en.wikipedia.org/wiki/List_of_pollen_sources

Let's continue beekeeping together!

https://beekeep.blue/join

The Beekeeping Blueprint Community

- Connect with experienced beekeepers around the world
- Share photos of your hive or harvest
- Swap tips and ideas
- Troubleshoot issues
- Keep learning!

JOIN HERE: https://beekeep.blue/join

Glossary

abscond: when an entire colony of bees leaves their hive altogether due to unfavorable conditions (lack of food or water, pests or disease, regular disturbances, changes in weather, bad ventilation, issues with the queen, etc.)

alcohol wash: a method used to assess and monitor the population of Varroa mites in a beehive by taking a sample of bees and washing them in an alcohol solution, causing the mites to detach from the bees and sink to the bottom of the container

apiarist: a person who keeps and manages honeybees; a beekeeper

arboretum: a place where a diverse collection of trees, shrubs and woody plants are cultivated for scientific and educational purposes

bearding: when bees cluster temporarily at the entrance or on the outer surface of the hive to create space inside the hive, allowing for better ventilation and heat dissipation

bee bread: a type of food made by honeybees, which is a vital source of nutrition containing proteins, vitamins, minerals and beneficial bacteria

bee brush: a tool used by beekeepers to gently move bees from one area to another without harming them

bee space: the specific distance (3/8 of an inch) maintained between the frames and other structures within the hive that allows honeybees to move and work efficiently

bottom board: a flat, rectangular board that serves as the base of the beehive, helping to protect from pests, retaining heat (solid bottom board) and encouraging airflow (screened bottom board)

burr comb: the irregular or unintended wax comb that honeybees build between frames or in unwanted spaces within a beehive

brood: the developing stages of honeybees within the colony, consisting of eggs, larvae and pupae

brood chamber: a box within the hive where the queen bee lays her eggs and where the majority of the brood is reared

candy board: a mixture of granulated sugar and water that has been set and hardened in a frame or mold, serving as a supplemental food source

capping scratcher (or uncapping tool): a narrow brush with sharp wire bristles used to scratch open wax cells to retrieve honey or uncap single cells to inspect larvae

checkerboard: to alternate frames of brood with frames of drawn comb/foundation, encouraging bees to expand their brood nest and store more honey

chemical identity: various chemical substances that play important roles in the biology, behavior and communication of bees

cluster: the behavior and formation of honeybees huddling closely together in a tightly knit group, particularly during colder weather, to maintain warmth and survive adverse conditions

colony: a structured social group of bees that live together in a cooperative system, consisting of three main castes of bees: the queen bee, worker bees, and drones

Colony Collapse Disorder (CCD): a phenomenon in which the majority of worker bees in a honeybee colony disappear or die, leaving behind a queen and a number of immature bees

cross comb: the result of bees building comb that is not aligned with the frames or foundation of a hive, connecting comb sections between adjacent frames

cull: to remove or eliminate certain bees or colonies in order to manage the overall health or productivity of the bee population

Deformed Wing Virus (DWV): a viral infection transmitted through the Varroa mite, causing wing deformities and reduced flight ability

drawing comb/foundation: the process of bees building their wax honeycomb

drone comb: specially designed frames or sections within a beehive that contain slightly larger cells and encourage the development of drone bees

emergency cell: a type of cell built in response to the sudden loss of the colony's queen, where normal brood cells are converted into supersedure cells by nurse bees

entomologist: a scientist who studies insects

entrance reducer: a narrow strip of wood or plastic that is placed at the hive entrance, partially reducing the opening through which bees can enter and exit and helping the colony defend itself against potential threats

escape board: a board with specialized holes or passages that allow bees to exit the honey super but make it difficult for them to re-enter (also known as a "bee escape")

established colony: a fully functioning bee colony that has already established itself within a hive

exoskeleton: a rigid external structure located outside the body that provides support, protection and attachment points for muscles and organs in certain animals

feeder: a device or container used to provide supplemental food to honeybee colonies when natural food sources are scarce or insufficient

forage: to collect nectar, pollen and other resources from the surrounding environment to sustain the hive

foundation: a panel made of beeswax or wax coated plastic that serves as a base and guide for bees to build their wax cells in a uniform, organized manner

frame: a removable rectangular structure typically made of wood or plastic that fits inside the boxes or supers of a beehive, which provides structure for bees to build their comb (it can have a foundation or be foundationless)

hive drifting: when bees from different colonies become disoriented or confused and end up entering a hive other than their own

hive tool: an essential tool used in beekeeping for manipulating hive components, inspecting beehives,and performing various hive management tasks

hygroscopic: the ability of a substance or material to attract and absorb moisture from the surrounding environment

inner cover: the part of a beehive (usually made of wood) that is positioned directly below the outer cover and above the topmost box or super, serving as an additional layer of protection and insulation for the hive while also allowing for ventilation and feeding access

introduction cage: a queen cage designed with mesh or perforated sides to allow for ventilation and interaction between the new queen and worker bees, facilitating the gradual acceptance of the queen in more aggressive colonies

mechanoreceptors: sensory structures that are responsible for detecting physical stimuli and movement, such as vibration

monoculture: the agricultural practice of growing a single crop species over a large area of land

mountain camp method: a supplemental feeding technique where a sheet of newspaper is placed directly on the brood frames within the hive and dry sugar is poured over top for the bees to access

mouse guard: a metal or plastic screen or grid that is placed over the hive entrance to guard against mice while allowing honeybees to come and go freely

nectar: sweet, sugary liquid produced by flowers, which is an important energy source for pollinators and is collected by bees to produce honey

nosemosis (nosema): a fungal infection that invades the intestinal tract of bees and disrupts their digestive system

nucleus colony (or nuc): a small-scale, self-contained honeybee colony created by beekeepers for various purposes such as queen rearing, colony expansion, hive management or bee breeding

open worker brood pheromone: a chemical substance created by uncapped worker bee brood that suppresses the functioning of worker bee ovaries and their ability to lay eggs

outer cover: the topmost part of a beehive (usually made of wood) that serves as the protective covering for the entire structure, providing shelter and insulation for the bees and their comb

overwintering: the process by which honeybee colonies survive the winter months, particularly in regions with cold climates where forage is scarce and temperatures drop below levels conducive to bee activity

oxalic acid dribble: a method used to control Varroa mite infestations, which involves applying a solution of oxalic acid to the bees and hive components—it's toxic to Varroa mites but has a minimal impact on honeybees when used correctly

packaged bees: a common method used in beekeeping to introduce a new honeybee colony into a hive, consisting of a group of worker bees and a mated queen bee, along with a small amount of syrup or sugar solution for nourishment during transport

pheromones: chemical substances produced and released by bees used for communication and maintaining the organization and functioning of the colony

pollen cleaner: a machine used to separate debris from bee pollen

pollen patty: a supplemental food source made from a mixture of pollen, protein supplements and other ingredients that promote brood rearing

pollen trap: a device installed at the hive entrance used to collect pollen from bees as they enter or exit the beehive, removing or scraping off the pollen pellets from their hind legs

pollinate: when pollen grains from the male part of a flower are transferred to the female part of the same or a different flower, leading to fertilization and subsequent seed and fruit production

proboscis: the long, straw-like tongue that bees use to drink nectar from flowers

propolis: a sticky resinous substance collected from various plant sources that is used by bees to seal and protect the hive

pupa cell: a specialized cell within the honeycomb where honeybee larvae undergo metamorphosis and transform into adult bees

queen catcher cage: a specialized tool consisting of a small cage (typically made of plastic or metal) designed to temporarily confine the queen bee for various purposes, such as hive inspections, queen marking, or queen replacement

queen cell: a specialized, more elongated cell created by worker bees to rear a new queen bee

queen cup: an extended beehive cell built as a potential site for rearing a new queen bee if necessary

queen excluder: a barrier that is typically placed between the brood chamber and the honey supers to prevent the queen from entering the latter while still allowing worker bees to pass through freely

queen pheromone: a combination of chemical substances released by the queen bee that influences the behavior of other bees in the colony

queenright: the state in which a beehive contains a healthy and actively laying queen bee

render: to process or extract the wax from honeycomb to obtain a purified form of beeswax that can then be used for various purposes

robbing: the aggressive behavior of honeybees from one colony attempting to steal honey or resources from another colony's hive, typically occurring when there is a scarcity of nectar or honey flow robbing screen: a barrier used to prevent robbing in a beehive that usually consists of a mesh or perforated material that allows for ventilation and airflow but prevents easy entry into the hive

royal jelly: a thick, creamy substance (made up of proteins, sugars, fats, vitamins and minerals) produced by worker bees that serves as the queen bee's main food source

slatted rack: a wooden frame with evenly spaced bars running across it that can be placed between the bottom board and lower brood box to increase airflow, helping to regulate hive temperature and humidity levels

smoker: a tool that consists of a metal canister with a bellows attached, allowing the beekeeper to blow air into the canister to create a controlled stream of smoke, which is used to calm honeybees during hive inspections and manipulations

spermatheca: specialized organ generally found in female insects (including the queen bee) that's responsible for storing and retaining sperm received during mating for later fertilization of eggs

sticky board: a flat, rectangular board that is placed below the bottom board, which acts as a trap for mites and other small insects

super: a section or box that is added to the hive structure above the brood chamber to provide additional space for bees to store honey

superorganism: a highly organized social unit composed of multiple entities, where members work together cooperatively and perform specific roles to support the overall functioning and survival of the colony

supersedure cell: a type of cell built around certain larvae when preparing to replace a colony's aging, injured or failing queen; these are usually found scattered across the face of frames

swarm: a large group of bees that leave their original colony in search of a new home

swarm cell: a specialized cell that is used to raise a new queen when the colony is preparing to swarm; it is larger than average, often built in clusters and found hanging vertically from the bottom of frames

tracheal mite: a parasitic mite that lives inside the tracheal tubes (breathing tubes) of honeybees, blocking their airways and feeding on the hemolymph (similar to blood), causing weakness

uncapping tool (or capping scratcher): a narrow brush with sharp wire bristles used to scratch open wax cells to retrieve honey or uncap single cells to inspect larvae

vaporization: a vapor or gas treatment used for pest control and disease management in honeybee colonies

Varroa mite: a small parasitic mite that feeds on the blood of honeybees and is considered one of the leading causes of honeybee population decline

wax capping: a thin layer of beeswax that worker bees use to seal cells in a beehive to prevent any contaminants from entering the cell and preserving stored honey or developing brood

windbreak: a physical barrier or structure that is strategically placed around beehives to protect them from strong winds

winter patty: a supplemental food source high in carbohydrates and low in protein and fat that promotes the health of bees and is designed to help colonies survive the winter

About the Author

Jason Chrisman is a farmer, beekeeper and video content creator in central Ohio. He leads a homesteading lifestyle with his wife, daughter and their American Bully, Ladybug. Jason welcomed bees into his family the day a swarm made its way into his front yard in 2009. Today, along with dozens of hives, he also rears grass-fed cattle (mostly Angus beef), pastured chickens, turkeys, free-range ducks and Norwegian goats.

Jason created his YouTube channel JC's Bees as a personal diary and was delighted to find a supportive community offering advice and encouragement. Today, JC's Bees has tens of thousands of subscribers and aims to help beekeeping newcomers get started with their first hives. Jason's goal is to encourage and inspire his viewers to develop and perfect their beekeeping skills. The support and appreciation he has felt through his channel have led directly to the creation of the *Beekeeping Blueprint* book and online community platform.

Acknowledgments

I'd like to thank my wife, Amber Hill, and daughter, Chelsea Chrisman, for putting up with me and all the bees. There were indeed times when the bees got in the way of family plans, and you both accepted that with grace. Mom and Dad (Susie and Rodney Chrisman), thank you for teaching me the value of working hard for what I believe in. To my sister, Janelle, and niece, Chloe Ketron, thank you for always supporting my goals, even if keeping bees did seem a bit crazy at first.

Chuck Marshall—the contractor I worked for all those years ago—thanks for spotting that first swarm in my oak tree. You always did have a sharp eye for things! To my neighbor, Timothy Hahn, I'm grateful for the bee box you provided when I caught that first swarm. Without it, I might still be waiting for my big beekeeping break. I also need to thank my distant cousin for being a mentor to me at the very beginning of my beekeeping journey. You taught me a lot!

Finally, to anyone who reached out and offered their beekeeping advice over the years: the world needs more people like you! To all my followers and supporters who keep me pushing forward, know that you are appreciated and valued. To Nathan, Joe, Phoebe, Alethea, Moch and the rest who helped turn this book into a reality, a heartfelt thank you.